VOLUME FIFTY NINE

ADVANCES IN GEOPHYSICS

VOLUME FIFTY NINE

ADVANCES IN GEOPHYSICS

Series Editor

CEDRIC SCHMELZBACH

Institute of Geophysics,
ETH Zurich,
Zurich, Switzerland

ELSEVIER

ACADEMIC PRESS

An imprint of Elsevier

Academic Press is an imprint of Elsevier
50 Hampshire Street, 5th Floor, Cambridge, MA 02139, United States
525 B Street, Suite 1650, San Diego, CA 92101, United States
The Boulevard, Langford Lane, Kidlington, Oxford OX5 1GB, United Kingdom
125 London Wall, London, EC2Y 5AS, United Kingdom

First edition 2018

ISBN: 978-0-12-815208-9
ISSN: 0065-2687

For information on all Academic Press publications
visit our website at https://www.elsevier.com/books-and-journals

Publisher: Zoe Kruze
Acquisition Editor: Jason Mitchell
Editorial Project Manager: Joanna Collett
Production Project Manager: James Selvam
Cover Designer: Mark Rogers

Typeset by SPi Global, India

Contents

Contributors

Guillaume Caumon
GeoRessources (UMR 7359), ENSG, Université de Lorraine, Vandoeuvre-lès-Nancy, France

Stewart Greenhalgh
Department of Geosciences, KFUPM, Dhahran, Saudi Arabia

Matthew Rutty
Petrosys, Adelaide, SA, Australia

Cedric Schmelzbach
Institute of Geophysics, ETH Zurich, Zurich, Switzerland

David Sollberger
Institute of Geophysics, ETH Zurich; Seismic Apparition, Zurich, Switzerland

Florian Wellmann
Computational Geoscience and Reservoir Engineering (CGRE), RWTH Aachen University, Aachen, Germany

Preface

This volume of *Advances in Geophysics* is dedicated to two reviews that are focused on geological modeling and related uncertainties as well as on seismic single-station wavefield polarization analysis.

Geological models are an important component of geophysical as well as geological investigations. Such models are representations of subsurface information resulting from the integration of multiple aspects of geological realism in the form of observations, information, and knowledge, cast in geometric representations of subsurface structures. Wellmann and Caumon review recent developments in geological modeling methods as well as concepts and methods to analyze, quantify, and communicate related uncertainties. They introduce formulations for geological model representation, interpolation, and uncertainty analysis methods with the aim to clarify similarities and differences in the diverse approaches that were developed in recent years. The chapter by Wellmann and Caumon provides an entry point to recent developments in geological modeling methods. Furthermore, the authors develop a framework to consider model uncertainties, which supports the integration of geological observations and knowledge into geophysical interpretation, modeling, and inversion.

The polarization properties of seismic wavefields recorded by triaxial (three-component) stations can be exploited for seismic event detection, direction finding, and wavefield filtering. Greenhalgh et al. provide a tutorial on extracting seismic wavefield polarization properties from the eigen-analysis of a (complex-valued) three-component covariance matrix. The advantage of this complex-valued polarization analysis is that different modes of polarization can be examined with the same methodology. Greenhalgh et al. discuss the extraction of various single-station polarization attributes from synthetic and field data sets and demonstrate the application of polarization measures for wavefield characterization and filtering.

Enjoy your reading of Volume 59 of *Advances in Geophysics*!

Cedric Schmelzbach, Editor

CHAPTER ONE

3-D Structural geological models: Concepts, methods, and uncertainties

Florian Wellmann[a,*], Guillaume Caumon[b]
[a]Computational Geoscience and Reservoir Engineering (CGRE), RWTH Aachen University, Aachen, Germany
[b]GeoRessources (UMR 7359), ENSG, Université de Lorraine, Vandoeuvre-lès-Nancy, France
*Corresponding author: e-mail address: wellmann@aices.rwth-aachen.de

Contents

Abstract

The Earth below ground is the subject of interest for many geophysical as well as geological investigations. Even though most practitioners would agree that all available information should be used in such an investigation, it is common practice that only

ISSN 0065-2687
https://doi.org/10.1016/bs.agph.2018.09.001

a part of geological and geophysical information is actually integrated in structural geological models. We believe that some reasons for this omission are (a) an incomplete picture of available geological modeling methods, and (b) the problem of the perceived static picture of an inflexible geological representation in an image or geological model.

With this work, we aim to contribute to the problem of subsurface interface detection through (a) the review of state-of-the-art geological modeling methods that allow the consideration of multiple aspects of geological realism in the form of observations, information, and knowledge, cast in geometric representations of subsurface structures, and (b) concepts and methods to analyze, quantify, and communicate related uncertainties in these models. We introduce a formulation for geological model representation and interpolation and uncertainty analysis methods with the aim to clarify similarities and differences in the diverse set of approaches that developed in recent years.

We hope that this chapter provides an entry point to recent developments in geological modeling methods, helps researchers in the field to better consider uncertainties, and supports the integration of geological observations and knowledge in geophysical interpretation, modeling and inverse approaches.

1. Introduction and preliminary considerations

The geological space in the subsurface contributes in essential parts to the anthroposphere, as it has a direct influence on mankind and is equally affected by our technological society. Geological processes create significant heterogeneities at multiple scales below the surface of the Earth. These heterogeneities are essential sources of wealth for mankind (e.g., natural resource deposits) but they are also a main driver of natural hazards such as earthquakes and volcanic eruptions. On the other hand, we influence and alter the subsurface geological space, for example, due to resource and water extraction, or engineering and geotechnical applications. Therefore, the evaluation of the spatial distribution of properties in the subsurface plays an essential role for a wide range of applications, from scientific investigations, to geotechnical aspects (Culshaw, 2005), hydrogeological investigations (Anderson, 1989), raw materials and hydrocarbons (Ringrose & Bentley, 2015), and finally policy and public hazard information.

Since the first geological map by William Smith more than 200 years ago, geoscientists have put significant efforts in understanding and visualizing how rocks are organized below the Earth surface, both for practical reasons and out of scientific curiosity to understand the world below our feet. This knowledge is now commonly encapsulated in *structural geological models*, often simply referred to as *geological models* or *geomodels* in this context, as a representation of geometric elements such as rock unit boundaries, faults, horizons, and intrusions on a scale of meters to kilometers. The goal of this paper is to present recent advances in the field of geological modeling, with interpolation concepts and algorithms that allow us to represent what we do know—and approaches to analyze and quantify uncertainties to clarify what we do not.

Geological models are intimately linked to geophysics, as they can be seen as a spatial representation of specific aspects of geological knowledge. In this context, they often provide the basis for a spatial parameterizations of geophysical forward models, included in geophysical processing and inversion frameworks. At the same time, geophysical observations and interpretations are often an important input to geological models, and geophysical measurements are commonly used to validate geological models, in combination with petrophysical models.

In our experience, the geometric viewpoint of geological models can provide a link between geological and geophysical observations with additional geological and physical information and knowledge, as many relevant aspects share the concept of discrete interfaces in the subsurface as a unifying element, if only in different forms and appearances. To further explain this idea, we will first describe concepts of interface detection in geophysics and geology and review related work. Subsequently, we will describe the current state of geological modeling methods, specifically with the focus on the potential integration into joint geological–geophysical investigations. Even in the best possible combination of information, uncertainties remain—and a substantial part of recent research addresses the investigation and quantification of uncertainties in geological models. The most relevant aspects are described below, followed by a discussion of the concepts and the major challenges, and an outlook to interesting paths for future work.

1.1 On geological realism considered in structural geological models

The essence of many subsurface investigations is to obtain an estimate of the true physical property value or class at any location in a region of the subsurface. As we yet have no possibility to directly measure this value exhaustively in space, we need to combine all available measurements, observations, and knowledge to obtain an informed estimate for it.

Of course, the spatial distribution of properties in the subsurface is not random—it is rather the result of the highly complex and long geological history that finally led to the state of the system that we observe in its present form. Unfortunately, we cannot model the entire evolution of the Earth in such a detail that we would be able to obtain a completely realistic picture of the spatial distribution of all relevant properties. Instead, we aim to formulate models that capture essential aspects of this evolution, which we deem relevant for the purpose and scale of a specific investigation. This specific consideration of geological knowledge can be considered as *geologic realism*, and an essential question for each subsurface investigation is which type of geological knowledge and how much geologic realism is required for a specific modeling aim and purpose?

In this chapter, we focus on geologic realism with respect to dominant structural geometric features. The motivation for the consideration of these features is best described with an image of a geological structure, provided in the following section.

1.1.1 A natural rock structure as a motivating example

We take here an example of a folded carbonate layer in the Pyrenees (Aragon, Spain), presented in Fig. 1A. The (simplified) geological story on the area associates sedimentary and tectonic processes (Alonso & Teixell, 1992): during the Upper Cretaceous, sandstone (Marboré Formation) was deposited on top of the Pyrenean basement rocks, and was then overlaid during Paleocene by dolomite and massive limestone beds. Layered carbonate beds were deposited during the lower Eocene, then a sudden rise of the relative sea level led to the deposition of deep-marine turbidites of the Hecho Group. This sea-level rise is explained by a fast subsidence related to the Pyrenean compression which provoked a flexure of the lithosphere on either sides of the axial (collision) zone. The North/South shortening then affected these sediments, which accommodated deformation by folding and faulting.

This history is inferred from observations by applying basic stratigraphic principles established in the 17th century (Steno, 1669): more recent strata are deposited roughly horizontally on top of older strata; variants stem from later geological events. Geological concepts have much evolved since then, but these essential principles are still applicable in general.

Clearly, we cannot capture the entire complex history quantitatively, including the formation of the basement, the details of erosion and sediment transport and marine life which lead to the formation of sediments, global and regional effects like sea-level rises and subsidence, and finally the complexities of the Pyrenean orogeny. However, what we *choose* to consider when we talk about structural geological models are *surfaces* in 3-D space, which capture the essence of geological events in this complicated and long

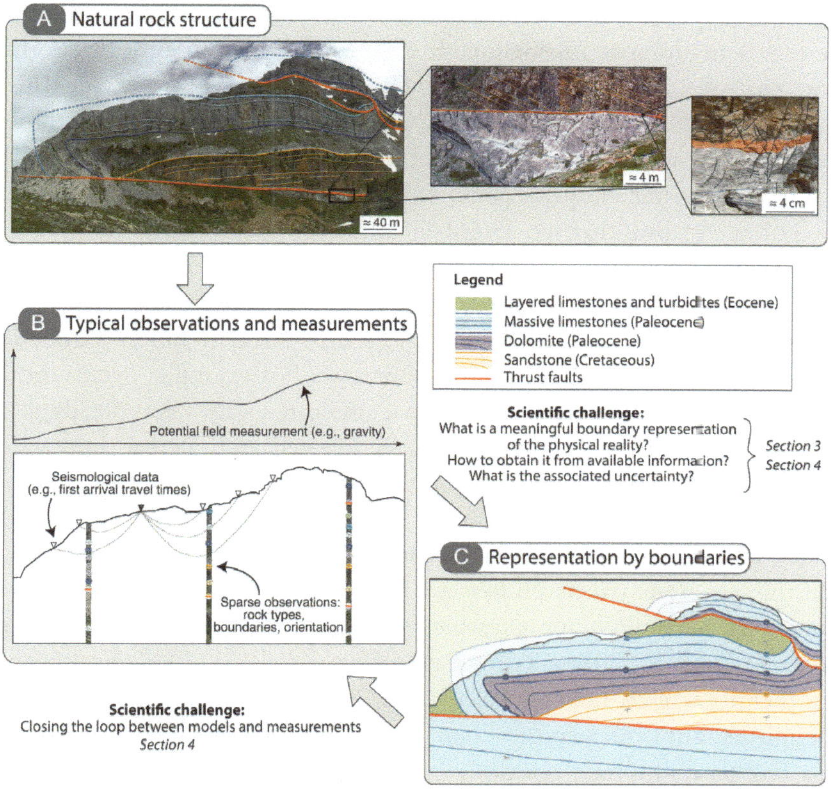

Fig. 1 Example of complex geological structures as seen on outcrop (A): this fault-related fold involves juxtaposition of sharp boundaries between turbidites, carbonates, and sandstone. As typical observations are incomplete and sparse (B), a challenge is to decide on which boundaries to represent, and to determine their geometry consistently with conceptual knowledge and the associated interpretations. Location: Aragüés del Puerto, Aragon, Spain.

history and which are often manifested at discrete points in time.[a] These surfaces are, for example, related to changes in the depositional system, leading to different rock types that we observe today, or related to deformation events like faulting or ductile deformation.[b]

If we now consider the realistic case that we cannot observe the geological domain in an outcrop in the Pyrenees, but that it may be several meters

[a] Discrete in the sense of very short, in a geological time frame.

[b] Note that this explanation is mostly for sedimentary systems, but equal considerations and thoughts can lead to structural models of poly-deformed magmatic and metamorphic rocks (e.g., Calcagno, Chiles, Courrioux, & Guillen, 2008; Maxelon, Renard, Courrioux, Brändli, & Mancktelow, 2009), or even unconsolidated rocks and soil systems.

to kilometers below the surface, then the next question is how this geometric representation can be obtained from only a limited amount of information that we typically have (Fig. 1B): boreholes may provide direct observations, but only at very limited points in space, and geophysical measurements provide additional indirect information. How can we "join the dots", considering all of this partial information, and equally use the aspect of geologic realism that we described above?

The fundamental goal for the theory of geological modeling is to obtain methods that allow a reasonable description of geometric and *topologic* features in full 3-D space (Fig. 1C). As all models, this description is bound to approximate the truth within some tolerance. For example, details on the right of Fig. 1A showing some internal structures, joints and details about the fault core zone, may be consciously ignored and deemed impractical to represent explicitly at the scale of interest. The construction of a geological model is limited by the ability of the selected methods to represent the level of geological realism that is considered on a specific scale. The geological models presented here often act as the *framework models* for models of finer-scale variability, for example, using geostatistical approaches (e.g., Pyrcz & Deutsch, 2014a; Ringrose & Bentley, 2015). Geological models are therefore often hierarchical models, combining multiple scales. Even though we limit our exposure here to structural models, we will mention this link at appropriate positions in the paper and highlight its relevance in the discussion.

Also, as our information and knowledge is limited, the description may be locally inaccurate. This is illustrated by some differences between the reality in Fig. 1A and the model in Fig. 1C: Surfaces are locally smoother in the model, and a small fault visible in the outcrop is missing in the model. As a combined result of limited precision, conceptual, and observation limitations, each geometric representation will be uncertain. An increasing amount of recent research is going into the evaluation of these related uncertainties.

1.1.2 A note on model complexity

An important aspect of geologic realism is the question of complexity of the investigated geological setting, and specifically, how much of this complexity should be represented in a subsequent geological model. We refer here with geological complexity purely to the complexity related to the geometric setting, in line with the purpose of this paper (see also Jessell et al., 2014;

Pellerin, Caumon, Julio, Mejia-Herrera, & Botella, 2015; Wellmann, Horowitz, Schill, & Regenauer-Lieb, 2010a, for more details on this consideration). Typical levels of complexity that we will consider are sketched in Fig. 2, and concern the number of geological features of interest (e.g., number of geological interfaces or contacts between interfaces), and the possible relationships between these features.

The simplest possible structural elements in geological models are continuous subhorizontal interfaces (Fig. 2A): at each location (x, y) in the model domain, this interface has exactly one corresponding z value as a function of this location, $z = f(x, y)$. We refer to these methods here as

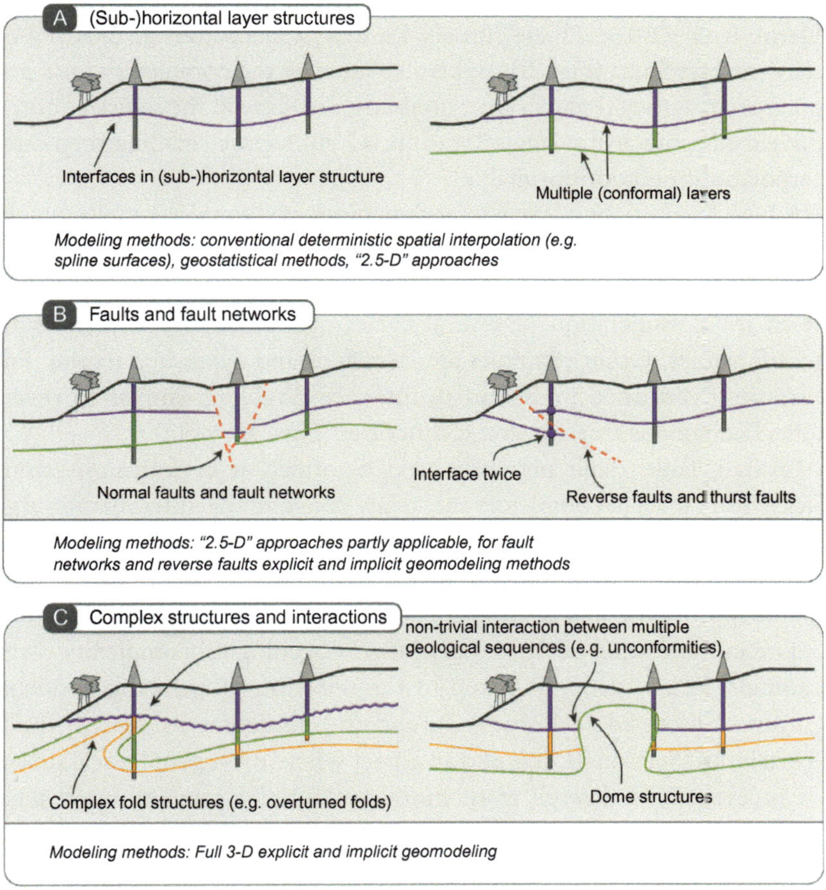

Fig. 2 Complexity from the viewpoint of structural geological modeling: different geological settings and possible geomodeling approaches.

map-based.[c] This type of structure can be modeled with a wide range of existing interpolation techniques, including geostatistical approaches. The situation is slightly more complex for multiple conformal layers (Fig. 2A, right), as the relationship between layers has to be considered (e.g., layers should not cross, constraints on thickness variations), but techniques for these elements also have long been available (e.g., Abrahamsen, 1993). More details are provided in Section 3.2.1.

The modeling situation is more complex when faults and fault networks are considered (Fig. 2B). First, faults add additional geometric elements that need to be modeled. Furthermore, geological continuity across faults often has to be taken into account, as faults act as discontinuities in the geometric sense and lead to a significant increase in topological complexity (e.g., Pellerin et al., 2015; Thiele, Jessell, Lindsay, Ogarko, et al., 2016). And finally, reverse faults (Fig. 2B, right) also increase the complexity for a geological layer, as they lead to a layer doubling, resulting in the interface potentially existing twice at a single location (x, y) in space, making map-based interpolation methods unsuitable.

A high level of complexity in both geology and topology is also quickly reached when more complex poly-deformed terranes and the interaction between geological sequences is taken into account (Fig. 2C), for example, due to the consideration of several cycles of sedimentary sequences and unconformities. Other examples are intrusions and dykes in a region. Furthermore, overturned folds lead to interface doubling, similar to reverse faults. The same is true for dome structures (Fig. 2C, right).

Overall, faults, fault networks, and deformation or intrusion settings quickly add a level of complexity that many conventional interpolation algorithms cannot consider and thus call for dedicated full 3-D geological modeling techniques. These methods will be further discussed in Section 3.1.3 for explicit approaches, and Section 3.2.3 for implicit approaches.

This consideration of the required level of geological complexity is very important, as it is strongly linked to the selection of a suitable modeling algorithm (discussed in Section 3), and the possibility to analyze model uncertainties (Section 4). It is also an aspect where most geophysical subsurface investigations diverge from more geologically oriented approaches. This distinction is explained in more detail in the following section.

[c] Similar in the level of complexity are extrusion (or "2.5-D") approaches which assume that geological volumes can be represented by extruding a surface along a direction.

1.2 Two viewpoints on interface detection

The problem of interface detection has occupied geophysicists and geologists for a long time. Arguably, two different viewpoints on this problem have emerged in this context: an approach to detect interfaces on the basis of geophysical data and measurements, described as the *data-driven* approach below, and an approach on the basis of geological observations and considerations, referred to as a *geomodel-driven* approach in the following. This contrast in the view on the same problem potentially derives from the academic background and language, and both approaches take different assumptions, but also share many commonalities.

The essential aspects of both viewpoints are presented in Fig. 3 and we will develop our interpretation on these viewpoints in the following and especially illustrate how they interact. For clarity, encircled numbers are added in the following text (e.g., ③) to refer to positions in Fig. 3. We aim to provide an overview of different aspects related to geological modeling and the link to geophysics, but also hope to provide a framework which can be helpful to place the great body of existing work, or new approaches in this field, into context, and to detect potentially interesting links to related work.

Both viewpoints initiate from the same object of interest: the physical reality ① of the subsurface. As we cannot directly measure this reality in a comprehensive way, we combine models of measurements of different types with mathematical and numerical models to finally arrive at an approximation of reality, ②. Both paths are linked in multiple ways and share several commonalities. They are both based on fundamental principles ③, for example, physical laws, and also on geological principles. In the following description, we aim to specifically examine the role of geological knowledge ④ and highlight the role of joint inversion methods ⑤ as a way to combine geological and geophysical information in a principled way.

1.2.1 The data-driven approach

We first describe the data-driven approach to obtain an approximation of reality on the basis of geophysical measurements ⑤. The commonly expressed aim is then to obtain a subsurface model ② of interfaces or directly property distributions from these data. A great body of work has been performed in this field, and multiple methods are now available and used in scientific investigations and practical applications with specific considerations for each geophysical measurement type. We will limit the following

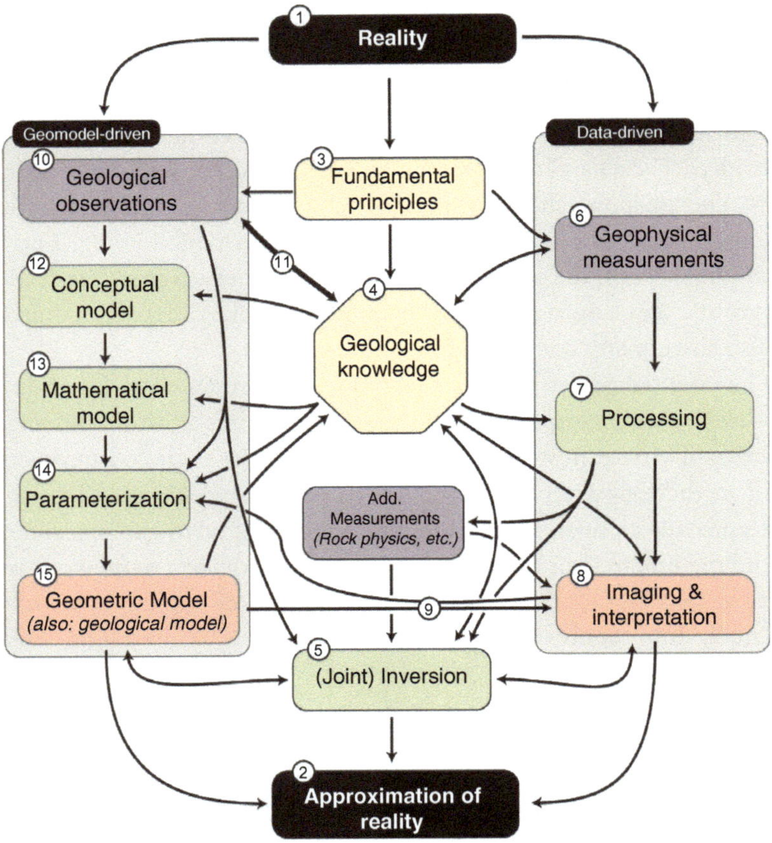

Fig. 3 Two viewpoints on subsurface interface and property detection; *blue boxes*: measurements and observations, *green boxes*: processing and modeling steps, *red boxes*: intermediate representations, *yellow boxes*: fundamental principles and knowledge. The *circled numbers* and *letters* refer to important aspects and links, explained in more detail in the text.

description to a general overview and provide some references to relevant literature with more details.

All geophysical data-driven approaches logically start with geophysical measurements ⑥. These measurement may have been taken for a specific purpose (as often the case in targeted exploration, for example), or as a general regional measurement (as, for example, magnetic and gravity state- or country-wide surveys). The measurements themselves are often derived from basic physical principles, but depending on the origin and type of measurement, geological knowledge ④ may have been considered for the measurements themselves, for example, in the design of a survey. A typical example is the decision about a suitable trace for a 2-D seismic or gravitational survey, perpendicular to the strike of the investigated geological structures (e.g., Telford, Geldart, & Sheriff, 2009; Yilmaz, 2001). At the same time, geophysical measurements can be used to derive first insights into the subsurface. An experienced applied geophysicist may be able to directly interpret measured signals, for example, raw gravity or magnetic measurements to obtain a first insight into subsurface structures. Geophysical measurements themselves can already contribute to a general geological knowledge of a specific region.

Still, the main contribution of geophysical measurements is usually obtained after dedicated data processing ⑦. Examples are the various steps in seismic processing from raw data processing and stacking, over migration, to time-depth conversion to obtain a reflection image (e.g., Yilmaz, 2001), or the various corrections applied to potential-field data (e.g., Jacoby & Smilde, 2009; Telford et al., 2009).

Geological knowledge may also enter the processing workflow at this stage, for example, in the form of bounds for velocity models for migration in seismic processing. An example for the use of geological reasoning in the analysis of potential-field data are also the various signal processing methods, which are often based on physical principles (for example, considering the upward continuation of wavelengths of gravity anomalies), but may also directly consider geological interfaces as the main object of interest, for example, when using wavelet-based methods (e.g., Hornby, Boschetti, & Horowitz, 1999).

In seismic processing and imaging (Yilmaz, 2001), the large redundancy present in reflection seismic data makes it possible to estimate a velocity model of the subsurface and to produce seismic images with limited geological information. In this case, the common approach is to use smoothness constraints in the regularization method; whereas it may be a suitable approach to consider smoothly varying velocities under the Born approximation, many aspects of geology cannot be considered in this way, because major and significant changes between subsurface areas with distinctively different properties appear on many sharp boundaries at multiple scales, as obvious in outcrops and rock-faces in nature (Fig. 1A).

From processed data, two paths can be taken: the first is a direct interpretation of the processed data ⑧, and the second a further combination with additional mathematical methods in an inversion framework ⑤. In the first case, the aim is to produce a geophysical image, for example, a profile of reflection seismic, or a spatially interpolated map of processed potential-field measurements at the surface, also then often called a geophysical map (e.g., Dentith & Mudge, 2014; Jacoby & Smilde, 2009). In this step, the geophysical measurements are necessarily combined with geological knowledge in order to derive a meaningful interpretation (e.g., Saltus & Blakely, 2011). In a minimal form, the concept of interfaces is used in the interpretation step, as obvious in the detection of interfaces on reflection seismic images (e.g., Yilmaz, 2001), or lineaments on gravity maps (e.g., Dentith & Mudge, 2014). However, there is also a significant contribution of more expressive forms of geological knowledge in this step. For example, in the case of seismic interpretation, Bond, Gibbs, Shipton, and Jones (2007) and Bond, Lunn, Shipton, and Lunn (2012) showed the significant influence of prior geological knowledge about a region on the interpretation of structures from seismic images. Even in the case of relatively well-constrained geological knowledge, uncertainties may exist in the location of the number, the geometry, and the connectivity of faults.

In addition to geological concepts, geological models often inform the interpretation step of geophysical data. The most obvious link is in the form of geophysical forward models, which are based on property distributions assigned to regions in a geological model, for example, a model of gravity response at the surface from defined mass distributions in the subsurface (e.g., Jacoby & Smilde, 2009).

Finally, interpreted images or geophysical maps are already often an endpoint in the data-driven workflow and taken as a suitable approximation of reality ② for a specific purpose. At the same time, the additional understanding during the intellectual process of geophysical image and map interpretation also feeds back into the general geological knowledge about a specific region.

We now consider the second path from processed geophysical measurements ④ to final models ⑥ using inversion approaches ⑤. We aim here to specifically outline the link of various inversion approaches to geological knowledge ③, specifically with the link to the geometric mode of interface detection.

Typical inputs to inversion approaches are based on either directly using the processed geophysical measurements ⑦, or already results from the imaging and interpretation step ⑧. A direct combination exists in the inversion of interfaces from gravity and magnetic data (e.g., Jacoby & Smilde, 2009; Silva, Medeiros, & Barbosa, 2001). The nonuniqueness of the inverse problem to determine density distributions from these measurements has been the content of early research in the field of geophysics (e.g., Parker, 1974), and subsequent work on suitable constraints for inversion (e.g., Li & Oldenburg, 1998). Geological knowledge ④, potentially also derived directly from geological measurements, has, for example, been considered in the form of dip measurements (e.g., Lelièvre, Oldenburg, & Williams, 2012; Li & Oldenburg, 2000). Many approaches also exist to invert density or velocity distributions or interface positions from a geological starting model (e.g., Autin, Scheck-Wenderoth, Götze, Reichert, & Marchal, 2016; Bosch, 1999; Bosch, Guillen, & Ledru, 2001; Fullagar, Pears, & McMonnies, 2008; Gradmann, Ebbing, & Fullea, 2013; Guillen, Calcagno, Courrioux, Joly, & Ledru, 2008; Haase, Ebbing, & Funck, 2017; Joly, Faure, Martelet, & Chen, 2009; Schmidt, Plonka, Götze, & Lahmeyer, 2011; Zheglova, Lelièvre, & Farquharson, 2018). The direct link to the geological model itself ⑨ is here very obvious.

Similar approaches exist for the inversion of electric and electromagnetic data, as well as seismic data (e.g., Clapp, Biondi, & Claerbout, 2004; Guiziou, Mallet, & Madariaga, 1996; Hauser et al., 2011; Hauser, Gunning, & Annetts, 2015) In fact, the relevance of boundaries has also long been realized as a central aspect of geophysical imaging techniques and a lot of work has gone in this direction since the pioneering work of Gjøystdal, Reinhardsen, and Åstebøl (1985). However, several studies explicitly suggested that geological insight can help seismic interpretations (e.g., Osypov et al., 2013; Rankey & Mitchell, 2003) and that, even when using modern seismic acquisition and processing practices, there is still ambiguity in velocity models that can lead to significant uncertainties in the true position of events in images. Very interesting is the also increasing research on full waveform inversion of seismic data (e.g., Kamei & Lumley, 2017; Zhu, Li, Fomel, Stadler, & Ghattas, 2016). These approaches can certainly be considered as dominantly data-driven, however, nonuniquenesses still remain and the inversion results depends on an initial (or baseline) velocity model.

As a summary of the data-driven viewpoint based on geophysical measurements ⑥ to obtain an approximation of reality ②, we see that a lot of work is aimed at a quantitative and, as much as possible, data-driven and mathematical analysis of the obtained measurement data. However, there is the fundamental problem that the measured geophysical signals provide only an indirect measure of the subsurface interfaces and properties, and an infinite-dimensional manifold of solutions exists (Backus & Gilbert, 1967; Parker, 1974). From a mathematical point of view, this ambiguity leads to ill-posed problems, and these have to be addressed with suitable regularization methods (e.g., Aster, Borchers, & Thurber, 2005; Tikhonov, 1963).

In fact, the ill-posed nature of geophysical problems lead to many of the mathematical developments in this field (e.g., Kabanikhin, 2008). In order to obtain nonunique and meaningful solutions, many decisions have to be taken, beginning from suitable regularization methods, to processing steps and the consideration of additional information in the form of geological models or constraints. There exists, therefore, no completely *objective* result for the inversion of geophysical data.

Many (if not all) of the steps in the "data-driven" approach therefore contain at least some aspects of geological knowledge. Several authors explicitly state that geological considerations increase the effectiveness of geophysical inversion (e.g., Fullagar et al., 2008). The fundamental question is therefore how this geological knowledge is obtained—and how relevant parts can be implemented at suitable steps in this "data-driven" viewpoint. These aspects are best explained in the framework of the "geomodel-driven" viewpoint, described below.

1.2.2 The geomodel-driven approach

A different point of view, anchored on centuries of work in the field of geological investigations, is to initiate the subsurface investigation from the viewpoint of main geological interfaces and the related primary information, as well as the associated geological concepts. This way of thinking is underlying the concept of geological maps and has subsequently been extended to full 3-D geological models. This viewpoint is at the core of understanding the relevant steps of geological model construction and the related uncertainties, and all essential aspects will be described in more detail in Sections 3 and 4 of this chapter.

In this viewpoint, primary data are geological observations and measurements, for example, in outcrops or wells ⑩. Here, the link between observations and geological knowledge ⑪ becomes essential and deserves a specific note: geological observations can never be separated from geological knowledge. Even taking a measurement of a surface orientation in an outcrop requires the previous geological knowledge about the interpretation of relevance for a feature, or the detection of a fault plane. This realization is at the core of the hermeneutic aspect of geology (e.g., Frodeman, 1995) and described in more detail below.

The set of observations, combined with the purpose of the model and the geological knowledge, typically first enter into a *conceptual model* ⑫. This conceptual model should consider all geological elements and available information, as well as aspects of geological knowledge with relevance for a specific study. This conceptual model is then the basis for the decision on an appropriate *mathematical model* for the geometrical interpolation in space ⑬. Typical interpolation functions and the associated assumptions and limitations are described in detail in Section 3. We note here that this decision is also based on the question of geological realism to be considered and, most importantly, on the level of complexity.

Once the representation and associated interpolation methods are defined in the form of a mathematical model, we can assign geological observations as parameters to this model ⑭. This step includes the definition of all relevant geological elements, as well as their interaction, for example, the number of layer interfaces, and the structure of a fault network. This decision is clearly influenced by the conceptual model, geological measurements, and the amount of available information, but, it is also often limited in practice by the specific mathematical model that is applied. The choice of a specific type of model representation and geological interpolation function determines which data can be used—and which cannot.

The most commonly used geological observation type is interface points (e.g., Dubrule, 1984; Hardy, 1971; Mallet, 1992), but the direct integration of orientation measurements is only possible in a few interpolation algorithms (e.g., Frank, Tertois, & Mallet, 2007; Hillier, Schetselaar, de Kemp, & Perron, 2014; Lajaunie, Courrioux, & Manuel, 1997). We specifically note that the set of geological input parameters can also contain drawings of interfaces in cross sections, generated by a geological expert, or more subjective knowledge conveyed through interpretive interface points. Yet again, other interpolation methods can consider thickness information (for example, kinematic algorithms, Jessell, 1981; Wellmann, Thiele, Lindsay, & Jessell, 2016).

Next to direct geological observations and measurements, interpreted geophysical images, and sections ⑧ are commonly used as an input to the parameterization of interpolation functions. Typical examples are reflection seismic data and geophysical maps of potential-field data. In many cases, this information enters directly into the interpolation step (e.g., interface picks in a seismic section), in other cases, it is included through the intermediate step of geological knowledge in a region, which is often informed by the consultation of all available geophysical images and interpretations and then considered in the definition of the conceptual model ⑫ and the choice of the mathematical model ⑬, as well as the parameterization of the interpolation functions ⑭.

In addition to the parameters based on geological and geophysical observations and measurements, each interpolation function requires specific parameters. These parameters are sometimes estimated on the basis of observations (e.g., parameters of covariance functions in geostatistical interpolations), but often based on a general expected behavior (e.g., smoothness of the interface). In addition, we often need to consider multiple interpolation functions to represent different geological sequences and structures, and then also have to define the interaction of these functions. More details on typically used mathematical interpolation functions and model parameterizations are presented in Section 3.

Once interpolation function(s) and parameters are defined, we obtain a geometric representation of the defined geological features in space. The most common example today are full 3-D visualizations, either of surfaces as boundary representations or volume models on discretized mesh structures (e.g., Caumon, Collon-Drouaillet, Le Carlier De Veslud, Viseur, & Sausse, 2009), and these 3-D representations are often referred to as *structural geological models* or simply *geological models*.[d] We note, however, that 2-D sections are also a common representation, also in the form of a *geological map*, as the intersection between a 3-D geometric model and a digital elevation model.

The geometric model ⑮ is then an important aspect of the required approximation of reality ②. Other common aspects include property fields with additional rock property models, possibly also in combination with further small-scale models of spatial heterogeneity to represent variability on a scale below the considered structures (see Fig. 1).

In any case, given all aspects considered before, it is already obvious that geological models can contain several *sources of uncertainty*, starting from the initial geological observations, over potentially vague and subjective aspects of geological knowledge, to the choice of the mathematical interpolation approach, its geometric and topological structure, and any additional interpolation parameters. In addition, we also observe that the diverse set of input data and the required choices in the model construction step also lead to different *types of uncertainty*, for example, uncertainties that are related to fixed choices made by the modeling expert in the definition of the conceptual model, different interpretations of features on a geophysical image, or measurement uncertainties in primary data. The treatment of these uncertainties is therefore closely linked to different steps in the *geomodel-driven viewpoint*

[d] The term *geological model* is, of course, highly ambiguous and should only be used when the context and the relationship to geometric models is clear.

and the respective choices in each step. The different sources and types of uncertainty are further discussed in Section 2.

It is clear from the description above that the geological model or map is, in essence, only a representation of one or several interpolation function(s) and the defined parameters, even though this aspect is often obscured by the fact that geological modeling methods are often hidden behind "black-box" implementations in commercial software packages. This understanding will, however, become essential for the treatment of diverse aspects of uncertainty, as many methods require the possibility for an automatic update of the geological model. From the previous description, it is evident that this level of automation is generally possible when the choices and settings in the mathematical model ⑬ and the parameterization ⑭ are made transparent.

This automation opens the way to an estimation of uncertainties in geological models. This is a key aspect of recent research in this field, and methods to estimate, quantify, and visualize the effect of these uncertainties on the subsequent geological models are treated in Section 4. An important point to mention here is that this automation allows for a tight integration of the geological modeling step into inversion approaches. This feature is relevant as it allows, for example, the consideration of geological observations and general aspects of geological knowledge that cannot directly be integrated into the geological (forward) modeling step. In addition, it provides the possibility for a tight integration with geophysical measurements, through the use of geological models in joint inverse approaches ⑤. The basis for this integration is typically a geophysical forward simulation on the basis of the spatial distribution of rock properties, obtained from the geological model.

1.2.3 Combining elements

The separation into two distinct approaches is clearly a simplification, but it captures the two main viewpoints on this problem and it is also, in our experience, the cause for a lot of misunderstanding between researchers trained in different fields. Both approaches have strengths and weaknesses. However, we specifically want to highlight here the links and the high connectivity between these different viewpoints, as described above and illustrated in Fig. 3.

Two important links exist between both viewpoints and can address the above concerns: (a) the central aspect of geological knowledge and (b) integrated inverse approaches. Both aspects are directly related: the underlying geological reality is the combining element between all types of investigations, and our knowledge about it is therefore a central aspect, whereas integrated inverse approaches include the technical possibilities to obtain a model of the reality under the consideration of all available and relevant information.

A central problem with geological knowledge, though, it that it is very difficult to express in a quantitative and objective way. This aspect is fundamentally related to the fact that geological observations themselves cannot be taken without a framework of knowledge, as already described above. We can instruct someone how to measure a planar feature with a geological compass, but if we place this person into an outcrop, he or she will be completely lost without the knowledge of how to identify a meaningful planar feature to measure.[e] This aspect is encapsulated in the characterization of geology as a hermeneutic science with a strong interpretative character (Frodeman, 1995) and it leads to the fact that many geological observations have a subjective character. This is very different with many geophysical measurements, which can technically be taken (although certainly not interpreted) without any knowledge about the subsurface, purely on the basis of the measurement technique itself. For example, gravity measurements can be performed by experienced technicians. This is certainly a generalization, but overall a significant difference between geological and geophysical measurements and observations.

To restate this aspect in our context of subsurface investigations: we *cannot* deny the ubiquity of interpretative and subjective aspects—but we *can* find ways to address this subjectivity. One approach addresses the elicitation of knowledge from experts as an attempt to reduce subjectivity in primary information (e.g., Curtis, 2012). A complementary approach is to find ways to reduce the effect of this uncertainty through the use of all available information—and to achieve this aim, we need to combine both viewpoints in the best possible way.

In the end, we obtain then the approximation of reality ②, informed by the geometric representation of the geological model and potentially optimized with additional geophysical information. As a result of the early integration of geological concepts and the tight link to geological knowledge, geomodels may include not only features directly visible on the geophysical

[e] To some extent, this is not true for all possible measurements. For example, it is possible to identify grain sizes on a core sample without knowing details about sedimentology. This is why geologists often have several terms to describe the same reality: some terms are mainly descriptive, while some others bear also an interpretation sense.

data at hand, but also features that are expected on the basis of related prior geological information (e.g., subseismic fracture networks). This concept is also commonly known as "Shared Earth Model" in the petroleum industry (Gawith & Gutteridge, 1996), or "Common Earth Model" in the mining industry (McGaughey, 2006a). Concrete examples of this approach to real-world problems have also been proposed in other fields such as geodynamics (e.g., Ziesch et al., 2015), seismology (e.g., Shaw et al., 2015) and geological engineering (e.g., Kaufmann & Martin, 2008).

The previous exposure does not answer the question of the required level of complexity for a specific study. As all modeling assumptions, this aspect requires the consideration of the specific model purpose. The geomodel-driven approach, therefore, should certainly not be confused with simply adding all detail and complexity that is present in nature, to obtain a "realistic" picture. The aim of modeling is always an abstraction, and this consideration holds here in the same way as in other cases (see also discussions in Ringrose and Bentley (2015) and Caumon (2018) and throughout this paper).

1.3 Typical input data

Three-dimensional geological modeling generally starts with spatial points, lines or surfaces interpreted from available observations and measurements. The data used in geomodeling is very heterogeneous and includes direct observations of rocks and indirect geophysical measurements. The heterogeneity of data is in itself a challenge in geomodeling: gathering all relevant information can take significant time due to accessibility and numerical storage issues. Another challenge is to assess the reliability of this information, as several sources of uncertainty affect these geoscientific data: position of the measurement, accuracy of the measurement device, volume being investigated. The quantity of interest (e.g., magnetic susceptibility) may also not directly relate to the feature to be modeled. As direct accessibility is limited to the Earth surface and to boreholes and galleries, spatial representativeness of Earth data is also very important. These limitations imply varying degrees of processing and interpretation even before three-dimensional modeling starts. Fig. 4 shows some typical types of earth data used in geomodeling, classified according to their volume of investigation and to the degree of interpretation they carry.

More precisely, geomodeling applications typically use several data types (e.g., Kaufmann & Martin, 2008; Kelk, 1992; Ringrose & Bentley, 2015):

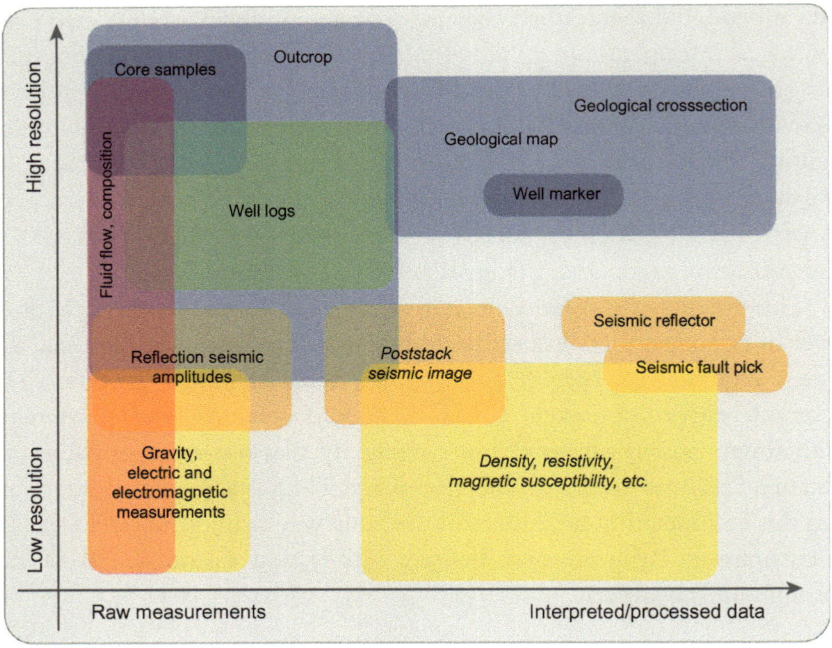

Fig. 4 Typical Earth data used in geomodeling. All data relate to a range of spatial resolution (vertical axis), and imply varying degrees of processing and interpretation (horizontal axis). Only a few data types (in *italic*) provide exhaustive spatial coverage in three dimensions. Colors highlight the type of data for various levels of processing and interpretation.

- *Surface data* include pictures and other remote sensing data (which may be combined into digital outcrop models), textual descriptions and drawings, georeferenced structural observations and measurements describing lithology, stratigraphy, unconformity, fault, stratigraphic orientation, fault striae, etc. These data are typically used directly as input for three-dimensional modeling, but are often initially translated into maps and cross sections.
- *Geological maps and cross sections* result from two-dimensional interpretations filling the gaps between surface observations. The reliability of this information essentially depends on the distance to actual observations and on the interpreter's skills, which are seldom objectively documented. In practice, 3-D visualization and modeling can help improving these interpretations by allowing the interpreter to consistently map in space.

- *Borehole data* include both well cores and geophysical well logs. Cores have the same features as surface data, but may be more difficult to interpret, as no rocks are generally observed away from the borehole. Geophysical logs are typically images or result from a local geophysical measurement which reflects physical properties in the neighborhood of the well. Images logs may contain significant structural information to identify layer dips and fractures. Overall, borehole data tend to be sparse and may have representativeness problems, as their location is often geared toward natural resources. The borehole data are generally interpreted in terms of location or range of location for the various geological surfaces.
- *Geophysical surveys* may either be directly interpreted or processed to provide a three-dimensional image. Reflection seismic data particular provide extremely useful information to identify reflectors and the associated discontinuities, which are then interpreted in terms of geological objects (e.g., Jacoby & Smilde, 2009). Depending on the type of the geophysical method, on the quality and on the spatial coverage of the data, interpretations may bear a variable value in resolving geophysical ambiguities.
- *Flow data* include flow rates at wells, transient pressure/ height and fluid composition measurements. This type of data does not explicitly include geological information, but can indirectly tell a lot about how subsurface reservoir rocks are connected.

Finally, if we neglect interpretation uncertainties, most geomodeling input data end up as labeled points indicating the most likely location and/or orientation of a particular geological surface. Additionally, we often have access to interval data (indicating the presence or absence of a particular geological feature in some region of space, a common case in geological field campaigns, for example) and trend data (indicating the slope, the fold axis or the dip and dip direction of geological interfaces). Additional information may also be used as input data, for instance the fault displacement or the thickness of a particular stratigraphic layer (Mallet, 1992), or may be considered as a way to (in)validate a generated model (e.g., De la Varga & Wellmann, 2016).

After the general overview of the considered aspects of geological realism, the conventional workflow with links to geophysics, and the typically used input data and the link to geological knowledge, we now consider the very opposite: the lack of precise knowledge about the subsurface, and the sources and types of uncertainty in geological models.

2. Classification and origin of uncertainties in geological models

Uncertainties have been an integral aspect of geological investigations since the early days of mining and exploration. This uncertainty is enshrined in common sayings between miners ("All is dark ahead of the pick"[f]), but it is just as well recognized for the oldest form of geological models, geological maps, where it has always been common to highlight uncertain areas and interface continuations, for example, using dashed lines or different line thicknesses and shading (e.g., Greenly, 1930).

A notable increase in research around uncertainties in geological maps and models occurred in relation to the development of digital mapping and modeling techniques. Early developments are represented in the work of Mann (1993), and the excellent review by Jones, McCaffrey, Wilson, and Holdsworth (2004) about uncertainties in geological mapping and modeling. In the latter paper, the authors describe the relevance of different types of geological knowledge (see also Section 1.2 and Fig. 3) and link these types of knowledge to different typical observations during geological mapping (see table 4 in Jones et al., 2004). These classification approaches were oriented along descriptive aspects, related to the process of data gathering and model building.

Similar considerations can be made for the typical workflow of geological model construction described above (Section 1.2.2). At each step in this modeling process, important decisions have to be taken, partly based on limited information and general aspects of geological knowledge, as well as measurements and observations, all subject to uncertainties. All of these decisions will manifest themselves in the final model (and, logically, also in the approximation of reality) in the form of model uncertainties.

In this section, we first describe sources of uncertainty related to each modeling step. We then provide an overview of uncertainties of typical input data. Finally, we describe approaches to classify uncertainties and propose a terminology based on uncertainty quantification concepts.

2.1 Uncertainty in the context of model construction

If we consider the workflow of model construction (Section 1.2.2, Fig. 3), then uncertainties enter at several levels: (a) at the definition of the

[f] German original: "Vor der Hacke ist es duster."

conceptual model and the decision of the mathematical model, (b) at the definition of the structure of the mathematical model, (c) when selecting parameter values for the mathematical model, and finally (d) at the exact value of the measurements. These steps and related uncertainties are presented in Fig. 5.

If we start from the model construction phase (Section 1.1), we first define a conceptual model, and here uncertainties are related to the lack of geological knowledge. This step is also heavily influenced by personal experience and belief (e.g., Bardossy & Fodor, 2004; Bond et al., 2007; Frodeman, 1995). On this basis, we decide on the mathematical model to perform the modeling step (described in more detail below, Section 3.2). We now need to decide on the structure of the model itself, for example, how many layers to include and how many faults. This step is closely related to the conceptual model, but we emphasize that it is not necessarily the same: a typical example would be the case where the exact number of layers is uncertain, and we would like to evaluate it from data. Examples will be given below. At this stage, the model space contains all of the models that are possible on the basis of the mathematical model and the defined structure. Finally, we consider the data itself and we only keep the models which actually explain the observed data, given the measurement uncertainties. This step is commonly called *conditioning* to the data. However, even after this step, many models are possible that explain the data in the range of measurement uncertainties.

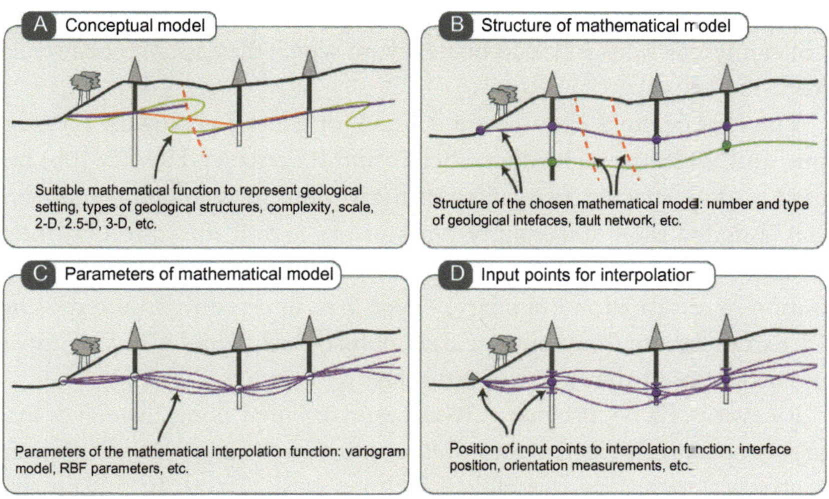

Fig. 5 Sources and types of uncertainty related to different modeling steps.

Several previous approaches attempted a classification of these sources of uncertainty into different types. In a general investigation of uncertainties in geology in the context of Nuclear Waste disposal, Mann (1993) promotes a classification of uncertainty types for geology following the previous (more general) work of Cox (1982). His basic types are:

- Type 1: Uncertainty due to observation error, bias and imprecision in the measurement process; could be reduced with additional measurements or better instrumentation;
- Type 2: Inherent variability and stochasticity; could be quantified with repeated observations (e.g., this includes long-term stochastic processes like earthquakes);
- Type 3: Ignorance, lack of knowledge, use of imperfect models and the need for generalizations.

He then describes the general sources of uncertainty and assigns these to the different types (see table 1 in Mann, 1993). He also makes the important point that a single *source* can be assigned to several *types* of uncertainty, which is already indicating that the separation of types is not completely clear. In his understanding, Type 1 and Type 2 are *quantitative* and can be described with probability density functions (pdf's) in a probabilistic framework, whereas he considers Type 3 as *qualitative* uncertainty, which could potentially be described with Fuzzy set theory (Zadeh, 1965) and also mentions the possibility for model comparison for Type 3 uncertainties. A similar separation into three types of uncertainties has been proposed by Dubois and Prade (2012), and Bardossy and Fodor (2004) provide an even more basic separation into uncertainties due to natural variability and uncertainties due to human imperfections and incompetency (to which they assign measurement errors, see below).

The classifications above were defined for general geological investigations, and Wellmann, Horowitz, Schill, and Regenauer-Lieb (2010b) proposed a more specific adaptation to the context of geological modeling. They assigned these uncertainty types in the logical structure of a model building process, where Type 1 uncertainties refer to input data (e.g., the position of a formation boundary), Type 2 to uncertainty in interpolation and extrapolation from known data points, and Type 3 to incomplete knowledge about the structural existence.

Jones et al. (2004) provide a classification oriented along "levels" of mapping campaigns, separated into *Data acquisition*, *Primary interpretation*, and *Compound interpretation* (see table 4 in Jones et al., 2004). In this list, they do not separate quantitative from qualitative uncertainty estimates, but they discuss the fact that uncertainties can partly be quantified directly from measurement errors (e.g., positional error taken from GPS) or obtained from repeated measurements, but that also highlight that uncertainty related to interpretation is more difficult to quantify. Interestingly, they include in their level of "Compound interpretation" the complete essence of uncertainty quantification in the typical question: "How can I quantify the uncertainty associated with this sophisticated interpretive model that I have slowly built up through a long iterative process of data collection and individual primary interpretations?" (Jones et al., 2004). A similar view to the one of Jones et al. (2004) has been taken by McCaffrey et al. (2005) in a review of digital geological mapping methods. Both works express the hope that future work in the field of digital mapping would allow a better tracking of uncertainties through the entire modeling process.

The uncertainties at different levels of geological mapping in Jones et al. (2004) are also reflected in the detailed description of uncertainties due to *human shortcomings, incompetency, or inadequate conditions* in Bardossy and Fodor (2004), where they distinguish problems during sampling (representativeness, sampling patterns, sampling density), insufficient lab measurements, nonmeasurable properties, and vague qualitative descriptions ("rare", "common", "frequent", etc.). The last sources of uncertainty in their list (6–10) refer are also especially relevant in our treatment of geological modeling, and this requires some more detailed thought. As discussed in Section 1.1, geological models are simplifications of reality and these simplifications and generalizations are sources of uncertainty. This includes all aspects related to the general background knowledge (see Fig. 3). These uncertainties also fall into the group of "beliefs" by Zimmermann (2000). Bardossy and Fodor (2004) describe that checking these generic (conceptual) models is *the* most difficult task (in geological model construction). The next sources of uncertainty are related to the choice and the correct application of the mathematical model, and, finally, uncertainties related to the conclusions drawn from a specific geological modeling investigation.

In practical applications, it may furthermore well be the case that uncertainties are apparently increased when more data are considered. This is especially the case when some data or concepts are in conflict (e.g., Zimmermann, 2000). However, this is also a strong indication that either the model itself is not suitable or that some information which is not relevant to the specific model is being used. This type of uncertainty is also referred to as *disinformation* in the hydrology literature (e.g., Beven, 2016).

2.2 Uncertainty of data and measurements

As described in the previous section, uncertainties enter at all stages of model construction (Fig. 5). Here, we provide a brief overview of typical sources of uncertainty related to input data to geological models as described in Section 1.3, suitable probability distributions, and methods of expert elicitation to pool judgment from multiple experts.

2.2.1 Uncertainties related to different data types

An immediate input to geological models are geological field observations and measurements, either from dedicated field campaigns, or derived from geological maps (Fig. 3). Sources of uncertainty in this context have partly been described in the literature (as described in the previous section, e.g., Bardossy & Fodor, 2004; Bowden, 2004a; Jones et al., 2004; Mann, 1993; Wellmann et al., 2010a). Especially the estimation of uncertainties in geological mapping, evaluated in Jones et al. (2004) and McCaffrey et al. (2005), has partly been described with uncertainties in subsequent 3-D geological models in mind.

LiDAR and UAV devices, as well as diverse types of remote sensing data, are now increasingly used to gather geological surface information and provide an unprecedented amount of information that can be expected to constrain geological structural models (Bilotti, Shaw, & Brennan, 2000; Bistacchi et al., 2015; Caumon, Gray, Antoine, & Titeux, 2013; Fernández, Muñoz, Arbués, Falivene, & Marzo, 2004; Vollgger & Cruden, 2016). Cawood, Bond, Howell, Butler, and Totake (2017) performed a detailed analysis comparing LiDAR, UAV, and conventional field observations and measurements with respect to accuracy and the effect on subsequent geological models. Although limited to a specific case study, this is an important step forward and we can expect to see more work in this direction in the future. Another highly active field of research is the use of remote sensing data to decipher geological surface information. For an introduction to the topic, see Lary, Alavi, Gandomi, and Walker (2016), the comprehensive work on Remote Predictive Mapping (RPM) by the Geological Survey of Canada (e.g., Harris, 2007; Harris et al., 2011; Schetselaar, Harris, Lynds, & de Kemp, 2007) and the interesting comparison by Cracknell and Reading (2013) and Cracknell and Reading (2014) for the specific aspect of geological mapping on the basis of remote sensing data.

Two aspects are receiving repeated attention in the discussion of uncertainties in geological interface points for geological mapping and modeling:

(a) data density and (b) geological complexity. An example for the effect of data density of interface points on a geological model is provided in the work of Putz, Stuwe, Jessell, and Calcagno (2006) for a model of a shear zone in the Eastern Alps. In a more recent work, Carmichael and Ailleres (2016) evaluate the effect of orientation data density on model construction and they provide a method for an automatic upscaling of orientation measurements for an optimized use in geological models. The second aspect of geological complexity is a lot more difficult to capture. Several studies have proposed measures of complexity (e.g., Ford & Blenkinsop, 2008; Zhao, Chen, Zuo, & Carranza, 2011), also partly with direct consideration of geological models (e.g., Lelliott, Cave, & Wealthall, 2009; Pellerin et al., 2015). Interestingly, Schweizer, Blum, and Butscher (2017) analyzed the effect of increasing model complexity on the estimation of model uncertainty. Also important to note is that geological complexity is closely related to geological history. However, there is yet no unifying measure of geological complexity to be directly used in a subsequent uncertainty study. This is certainly an interesting aspect of future research (see Section 5).

Other typical input data are borehole data, either in the form of well cores or geophysical well logs and the interpretation of geophysical wireline logs is a major topic in itself. At this point, we specifically want to mention quantitative approaches based on statistical rock physics Mavko and Mukerji (1998) and Mukerji, Avseth, Mavko, Takahashi, and Gonzalez (2001), as they have been applied in automatic extraction of geological attributes from well logs (e.g., Eidsvik, Mukerji, & Switzer, 2004), in recent studies also with an estimation of uncertainties (e.g., Grana, 2018; Grana, Pirrone, & Mukerji, 2012).

Probably the most dominantly used geophysical data in geological models are reflection seismic data, either classically in the form of 2-D seismic sections, but increasingly also in the form of 3-D seismic cubes (Dorn, 1998; Jackson & Kane, 2012) and it can be argued that reflection seismic data sets provide the closest form of a direct "imaging" of the subsurface structures (Davies et al., 2004). Uncertainties in seismic data have been analyzed in numerous studies (e.g., Glinsky et al., 2005; Suzuki, Caumon, & Caers, 2008; Thore, Shtuka, Lecour, Ait-Ettajer, & Cognot, 2002; Weinzierl, Schulte, & Aarre, 2016) and can mainly be assigned to processing steps (stacking, migration, time-depth-conversion) and the subsequent interpretation step.

A significant body of the early work on uncertainties in geological models has been linked to uncertainties in seismic processing, specifically time-depth-conversion (e.g., Abrahamsen, 1993; Abrahamsen, Omre, & Lia, 1991; Samson, Dubrule, & Euler, 1996). The advantage of this source of uncertainty is that prior knowledge exists in the form of expected functional relationships between time and depth domain, as well as the expected seismic velocities.

The relevance of this type of uncertainty is also reflected in more recent work. Li, Caers, and Sava (2015) developed a workflow combining geostatistical modeling and seismic imaging to asses the effect of velocity model uncertainty on generated images. The authors generate multiple image realizations on the basis of randomized velocity models. Interesting is then the link to subsequent seismic interpretations. The authors describe the adaptation of an image registration approach (the Thirion demon algorithm, Thirion, 1998) to track a manual interpretation of one seismic image (performed by a seismic interpreter) to all other randomly generated image realizations. Osypov et al. (2013) also highlight the impact of uncertainties in seismic velocity models by re-migrating horizon interpretations.

Recent work has also focused on the effect of interpretation uncertainties, mainly related to the interpretation of seismic measurements, but also for potential-field data. Uncertainties in seismic interpretation has received increasing attention in recent years (e.g., Alcalde, Bond, Johnson, et al., 2017; Alcalde, Bond, & Randle, 2017; Bond et al., 2007, 2012) and it is well recognized that seismic interpretations are strongly linked to the experience and knowledge of the expert (an aspect that goes hand in hand with the hermeneutic aspect of geology, see Frodeman, 1995, and Section 5). The studies also found a significant influence of expertise in specific tectonic settings, as well as the specific techniques used in the process itself, on the subsequent seismic interpretation, partly leading to biased interpretation. Studies using borehole data revealed similar patterns (Bond, Johnson, & Ellis, 2015).

As these interpretations are often a primary source of input data to geological models, their uncertainty can be considered as highly relevant to the subsequent geological model uncertainty. However, even for practices where reservoir properties are directly estimated from seismic data, studies show that these workflows are still affected by the uncertainty of the primary structural interpretation (Rankey & Mitchell, 2003). Interpretation uncertainty is also not resolved yet by autopicking algorithms (e.g., Herron, 2015), although increasing development on the basis of machine learning algorithms will potentially lead to improved algorithms in the future. Nonetheless, to manage possibly different interpretations and get a sense about how they quantitatively honor the initial seismic data, several authors have proposed to consider alternative geological structural models and then either generate synthetic seismic images by convolution (e.g., Botter et al., 2016; Lallier et al., 2012) or forward model wave propagation to generate synthetic seismograms (Irakarama, Cupillard, Caumon, & Sava, 2017). Computing misfit functions remains, however, a significant challenge as cycle skipping is likely to occur in the presence of large interpretation uncertainties owing for instance to poor illumination.

Gravity and magnetic measurements are commonly used geophysical measurements for geological models, both as input to geological models (derived from interpreted geophysical maps), as well as conditions for joint geological and geophysical inversion. A detailed treatment for the interpretation of gravity data and the related uncertainties is provided in Jacoby and Smilde (2009). Also here, recent methodological developments focus on diverse machine learning and optimization approaches with an additional assessment of uncertainties (e.g., Pallero, Fernández-Martínez, Bonvalot, & Fudym, 2015, 2017). Sources of uncertainty in specific case of aeromagnetic data are, for example, discussed in Holden et al. (2012) and Aitken, Holden, and Dentith (2013). These authors combine estimates of geological data quality, density, and interpretability, as well as orientation uncertainty to derive a measure of confidence in aeromagnetic interpretations. They also express the possibility to use these maps to obtain weighting factors for subsequent geophysical inversions.

2.2.2 Probability distributions for various sources of uncertainties

The various sources of uncertainty related to different input data types need to be formalized in some form to be used in a subsequent modeling study. This aspect alone can be difficult, as geological information often contains subjective elements due to the interpretative character of geological observations (e.g., Frodeman, 1995; Wood & Curtis, 2004). If uncertainties *can* be quantified, then the most common description is in the form of probability distributions, both when relating these uncertainties to outcomes in random experiments, as well as for a description of degrees of belief (e.g., Gelman, Carlin, Stern, & Rubin, 1995; MacKay, 2003; Tarantola, 2006).

Probability distributions assigned to the input data and parameters have a direct effect on the subsequent step of uncertainty quantification. Ideally, these distributions should be derived from the primary input data (e.g., device-specific measurement uncertainties, uncertainties in time-depth conversion in seismic images or from the segmentation of wireline logs) or from repeated measurements (e.g., orientation measurements of planar geological features). In this case, a probability distribution can then directly be obtained on the basis of this information, possibly also from subsequent processing steps, such as probabilistic plane fits to structural lineaments (e.g., Seers & Hodgetts, 2016).

For the cases where more subjective information about uncertainty is included, different types of distributions may be suitable. For a general overview of suitable distributions, see MacKay (2003), Gelman et al. (1995), and Tarantola (2005). An overview for the case of geological uncertainties is also provided in Wellmann et al. (2010a). Some sources of geological uncertainty can reasonably be described with a normal distribution, for example, when a position of an interface cannot exactly be determined due to a gradual transformation (e.g., in wireline logs). Another common case in geological mapping studies is that only outcrops are available where specific lithological units can be determined, but where the contact is not visible. In these cases, a uniform distribution could be applicable, when no other information about the contact is present. However, it should be noted that uniform distributions to restrict ranges or bounds should generally be avoided for high-dimensional cases, as uniform samples in a hypercube will be concentrated at the edges (e.g., Curtis & Lomax, 2001). When sufficient data are present and deemed representative, nonparametric distributions (i.e., histograms) may also be used, possibly after a distribution smoothing step.

For the specific case of orientation measurements, distributions should be assigned according to a suitable directional statistical distribution from the family of Fisher–Bingham distributions (Fisher, Embelton, & Lewis, 1993; Fisher, Comber, & Wadsworth, 2006; Fisher, 1953). The analog to a normal distribution on a sphere is the von Mises–Fisher distribution, which is appropriate for orientations as spherical vector data (Davis, 2002; Pakyuz-Charrier, Lindsay, Ogarko, Giraud, & Jessell, 2018). Uncertainties related to parameters of the applied mathematical interpolation model can potentially be derived when additional data is available (e.g., Aug, Chilès, Courrioux, & Lajaunie, 2005; Gonçalves, Kumaira, & Guadagnin, 2017).

2.2.3 Expert elicitation

Geological prior information is often subjective (e.g., Wood & Curtis, 2004). This subjectivity should not be seen as an impediment to the use of quantitative methods, but rather as a motivation to develop suitable methods to reduce the effect of subjectivity, as appropriately described by Curtis (2012). One important aspect of subjectivity is the influence of prior experience and expectation, for example, in the interpretation of seismic sections (e.g., Bond et al., 2012; Rankey & Mitchell, 2003). Especially in the case of interpretations of uncertainty, cognitive biases such as overconfidence, anchoring, and availability bias can play an important role (Tversky & Kahneman, 1974).

Even considering the potential methods to consider uncertainties as input to geological models, it is important to stress that unfortunately not all knowledge can easily be described formally, let alone in a quantitative manner (See also discussion in Section 3.2.4. The difficulty to communicate knowledge is often expressed in the form of different stages of personal knowledge, following the pioneering work of Polanyi (1958):

- *Explicit Knowledge*: knowledge, which is readily available and communicated, e.g., accessible in digital form in tables or data sets;
- *Implicit Knowledge*: in the mind of an expert (or, generally, any person) and knowledge that *could* be made explicit and communicated, given time and effort;
- *Tacit knowledge*: this is the difficult aspect of knowledge, which exists in the mind of a person, cannot be easily communicated and made explicit; sometimes, this is what people would consider "gut feeling."

One method to make personal knowledge better available, and to address issues of bias and to reduce subjectivity are structured elicitation processes, going back to the work of Kidd (1987) and Ford and Sterman (1998). These methods have since been applied in geosciences (e.g., Curtis & Wood, 2004), also specifically with the aim to obtain probabilistic information from geological interpretations (e.g., Baddeley, Curtis, & Wood, 2004; Bowden, 2004b; Cooke, 1991; Polson & Curtis, 2010). More recent interesting approaches also include applications of expert elicitation to estimate subjective probabilities of natural hazards (Aspinall, 2010; Runge, Scherbaum, Curtis, & Riggelsen, 2013), where also here the goal to include quantified uncertainties in the decision process is clearly stated.

3. Geological modeling methods

Three-dimensional geomodeling is about modeling the topology (connectivity), the geometry and the properties of a particular region of the Earth (Mallet, 2002). In general, this process does not require to reconstruct the history of the area through time, but rather to interpolate between sample data while following some essential rules and principles. This lack of consideration for geological processes can be considered as a weakness: what is the scientific value of interpolation as compared to process-based models that try to explicitly reproduce coupled processes at geological time scales?

A first element of answer lies in the extreme difficulty to find appropriate physical laws and boundary conditions in process-based models that run over geological time scales. Honoring spatial observations with these methods calls for solving an extremely ill-posed inverse problem. Interpolation appears, therefore, as a viable and computationally efficient alternative to obtain a consistent picture of the subsurface while honoring a large variety of observations (Section 1.3). More philosophically, geomodeling methods form a very good example of the "distinctive set of logical procedures" discussed by Frodeman (1995) to describe geological reasoning. We fully agree with Frodeman that interpretation and historical considerations are essential to geological science in general and to geomodeling in particular.

In practice, geological history and interpretive concepts have been translated into geomodeling as: (1) ways to define and represent geological formations mathematically and numerically, and (2) workflows to build models honoring available information. In this section, we start by describing the typical input data used to create geomodels. We then we discuss possible methods to represent geological models, trying to establish the common mathematical approaches to the problem and how they relate to geological concepts. We then review the existing ways to interpolate structures from available data, before discussing validation strategies.

3.1 Geological models: Essential characteristics and concepts

3.1.1 Geomodel representations

The goal of this section is to describe how complex geological domains can be described in mathematical and numerical terms. Geomodels in the literature are generally described in terms of processes and modeling workflows using various types of meshes, grids, or meshless methods. To address subsurface uncertainty and discuss how geological concepts can be translated into mathematical terms, we think it useful to use a more abstract description, which we describe in the following. We will then review existing implementations in Section 3.2. In this section, we build on the same formalism as Sambridge, Bodin, Gallagher, and Tkalcic (2013) and further explain and elaborate the ideas expressed by Caumon (2018).

Fig. 6 shows various geomodels representing possible seismic wave velocity fields in the subsurface, and a corresponding set of one-dimensional profiles showing the true field (in red) and how the various representations

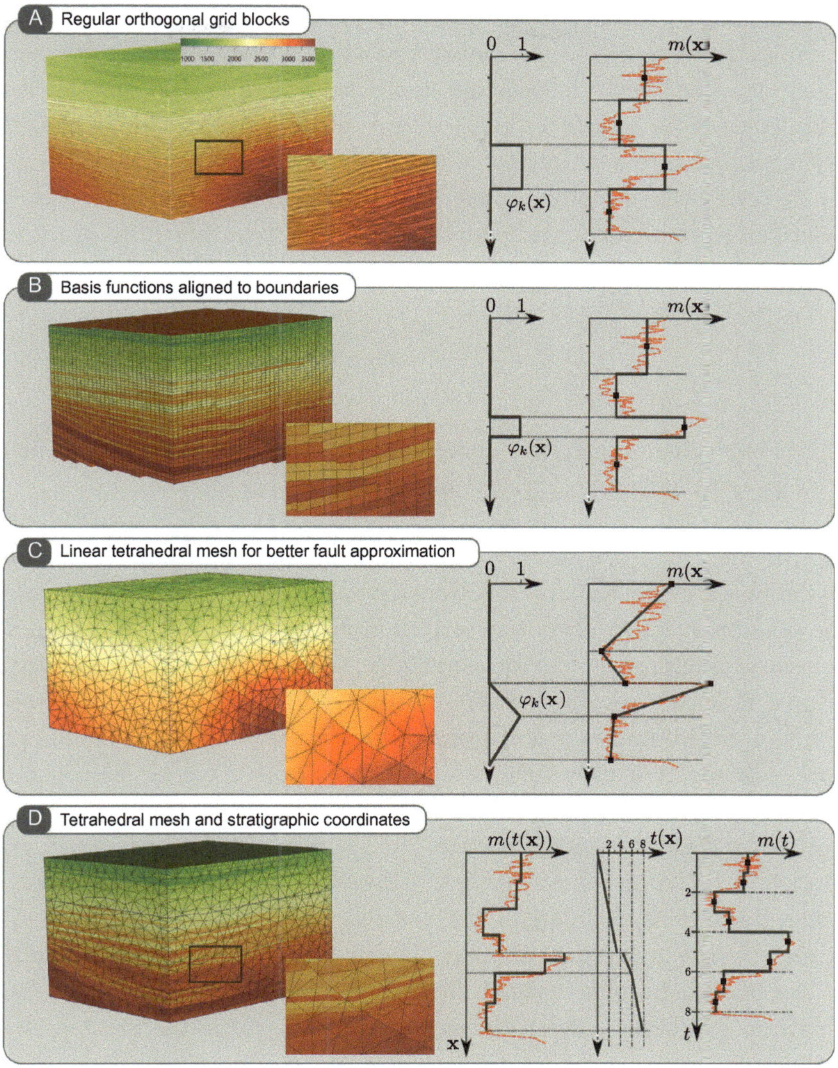

Fig. 6 Various representations of the p-wave velocity in a geological model. These models approximate the same structural geometry and show different seismic velocity models interpolated in space according to different representations. The right-hand side column shows some one-dimensional synthetic velocity profile (in *red*) and the model approximations (in *black*) corresponding to the left-hand side model (see text for more details).

approximate it (in black). These various options were all generated from sparse measurements and interpolated given the same structural geometry. They illustrate the basic principle used in numerical solid and physical modeling, which is to decompose the area of interest into a boundary representation or into a structured or unstructured mesh (Mäntylä, 1988).

In computational physics, all these geomodel representations can be described mathematically as a combination of K basis functions φ_k of the form

$$m(\mathbf{x}) = \sum_{k=1}^{K} m_k \, \varphi_k(\mathbf{x}) \tag{1}$$

to obtain a value of interest m at all possible locations $\mathbf{x}$ in a model domain (i.e., an approximation of reality, in the sense of Fig. 3).

In Fig. 6, m is the velocity field, K is the number of nodes (or grid blocks) bearing the static scalar velocity values m_k at the nodes (or grid blocks), and φ_k denotes the interpolation functions on the grid. The functions φ_k define the *geometric* structure of the parameterization. There is presently no general consensus about the type of functions $\varphi_k(\mathbf{x})$ to be used to best reflect what is known about the domain of interest. Most works choose a particular representation based on implementation comfort or on physical principles. The goal of this section and of Fig. 6 is to discuss about the adequacy of some choices to represent geological media.

For example, Fig. 6A discretizes the space by regular orthogonal grid blocks which correspond to piecewise constant basis functions. It overlays a fine decomposition representing the reflection seismic amplitudes to a coarser mesh representing velocities. The structural geometries in all models were obtained by expert-based interpretations if the seismic image. This Cartesian grid representation is very simple and widely used to represent geomodels in may fields, for instance in mining applications (e.g., McGaughey, 2006b). Depending on the grid resolution, this representation can approximate layer boundaries and smear petrophysical contrasts. In Fig. 6B, the basis functions are stretched or squeezed vertically so that they align on layer boundaries. This corresponds to a classical modeling choice made in reservoir modeling, corresponding to corner-point stratigraphic grids (Farmer, 2005). However, it is clearly visible that faults are approximated by stair-steps in this model representation.

Fig. 6C shows another option where a linear tetrahedral mesh is used to better approximate fault geometry. The tetrahedra are conformable to the fault surfaces, but cross horizons. The velocity field is linearly interpolated between tetrahedral nodes.

Once the basis functions have been chosen, the unknown coefficients m_k can be determined. Depending on the model purpose, m and m_k may represent one or several petrophysical parameters such as electric conductivity, seismic velocity, Lamé parameters, rock facies, rock composition, etc. Geostatistics is an essential framework to model the values m_k everywhere in space from available measurements. As very good reviews and textbooks on geostatistics already exist (Bosch, Mukerji, & Gonzalez, 2010; Bourges, Mari, & Jeannée, 2012; Chiles & Delfiner, 2012; Hu & Chugunova, 2008; Pyrcz & Deutsch, 2014b), we only stress here the basic idea: all the unknown parameters m_k are treated as correlated random variables. This general framework allows to describe spatial correlations for a given variable, and both colocated and spatial correlations between several petrophysical variables. Geostatistics also provides a framework to deal with the support (or volume) associated to each variable (Chiles & Delfiner, 2012; Journel & Huijbregts, 1978), but it is not widely used outside the field of mining geostatistics.

To describe complex domains, geomodeling generally follows the classical divide-and-conquer approach that is typically used in geological mapping. The domain is first subdivided into rock units that have some historical significance. Each unit may then be further subdivided into subregions that describe smaller features. Finally, the domain of interest is subdivided into a set of J regions $\{R_1, \ldots, R_J\}$. The physical properties of interest can be represented within each unit (or subunit) R_j by a set of basis functions $\varphi^j_k(\mathbf{x})$, $k = \{1, \ldots, K_j\}$ whose support covers and is completely included in R_j. The descriptions of the values of interest everywhere are then given by:

$$m(\mathbf{x}) = \sum_{j=1}^{J} \sum_{k=1}^{K} m_k \, \varphi^j_k(\mathbf{x}). \tag{2}$$

This type of representation is used for instance in the boundary element method, where an analytical petrophysical description is used within each subdomain. This type of approach has also been used a lot in geophysics for velocity modeling and ray tracing (e.g., Gjøystdal et al., 1985; Guiziou et al., 1996) or for potential-field modeling and inversion (e.g., Jacoby & Smilde, 2009; Wang & Hansen, 1990). As borehole measurements generally highlight that heterogeneities within regions cannot be modeled analytically, a discrete representation of values is often used, as shown in the models of Fig. 6. In practice, the region boundary needs to be discretized on the model mesh (e.g., Fig. 6A), or the model mesh is designed to conform to the region boundary, so that the mesh is aligned to layers (Fig. 6B) or faults (Fig. 6C).

As region boundaries often correspond to different rock types, it is common to extrapolate observed petrophysical values by treating region boundaries as discontinuities. This explains why sharp changes can be seen across horizons and faults in Fig. 6A, B, and D and across faults in Fig. 6C.

It can also be useful to represent each region by a an indicator variable $\mathbb{1}_j$, equal to 1 inside R_j and to 0 outside. Eq. (1) is then modified into:

$$m(\mathbf{x}) = \sum_{j=1}^{J} \mathbb{1}_j(\mathbf{x}) \left[\sum_{k=1}^{K} m_k \varphi_k(\mathbf{x}) \right]. \quad (3)$$

The indicator variable $\mathbb{1}_j(\mathbf{x})$, is itself a solid model which can be represented by the boundary surfaces of the region j or some other solid modeling technique (e.g., Mäntylä, 1988). Fig. 7 illustrates this on the two-dimensional example of a horizon H eroded by an erosion curve E. Both curves separate the 2-D domain into four subdomains E^+, E^-, H^+, and H^- described by four binary indicator functions $\mathbb{1}_{E^+}$, $\mathbb{1}_{E^-}$, $\mathbb{1}_{H^+}$, and $\mathbb{1}_{H^-}$, respectively. These functions are equal to 1 in their respective domain and to zero elsewhere.

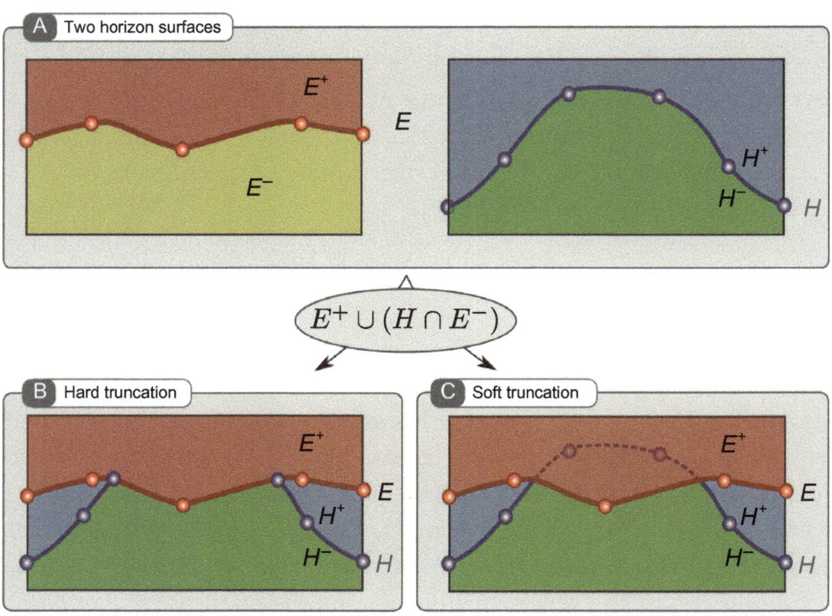

Fig. 7 Two possible ways to truncate a geological domain.

The erosion of the horizon curve H corresponds to a Boolean operation stating that no point of H, H^+ and H^- may exist in the region E^+ above E. Let us denote the noneroded horizon H as a linear combination of I basis functions χ_i^H (Fig. 7A, right):

$$H = \mathbf{x} \in \mathbb{R}^3 | \sum_{i=1}^{I} c_i \chi_i^H(\mathbf{x}). \tag{4}$$

In this example, the basis functions χ_i^H may for instance correspond to cubic Hermite polynomials centered on the curve nodes.

A first option to implement erosion is to explicitly trim the horizon H by the erosion curve E. As in Eq. (2)), this amounts to reparameterizing H by I' conformable basis functions $\chi_i^{H'}$ (Fig. 7B):

$$H_{erod} = \mathbf{x} \in \mathbb{R}^2 | \sum_{i=1}^{I'} c_i \chi_i^{H'}(\mathbf{x}). \tag{5}$$

Alternatively, eroding the horizon H may be achieved by canceling the Horizon H outside of the half-space E^- below the unconformity E. As in Eq. (3), this amounts to multiplying Eq. (4) by the indicator function $\mathbb{1}_{E^-}$ (Fig. 7C):

$$H_{erod} = \mathbf{x} \in \mathbb{R}^2 | \sum_{i=1}^{I} \mathbb{1}_{E^-} c_i \chi_i^H(\mathbf{x}). \tag{6}$$

The same rules can be used to account for all geological discontinuities, as will be further explained in Sections 3.1.2 and 3.1.3.

As in the above two-dimensional example, all the models shown in Fig. 6 started from a boundary representation of geological interfaces. Each variant corresponds to different choices for approximating the geometry of these boundaries: faces separating the blocks of a Cartesian grid (a) or of a curvilinear grid (b), or faces of tetrahedra aligned on the faults (c and d). As compared to (2), this formulation does not necessarily require to use a mesh conformal to the region boundary. This makes it easier to manage geometric changes of the region boundaries. This is one of the principles used for instance in the extended finite element method (XFEM) to model the growth of a propagating fracture (Moës, Gravouil, & Belytschko, 2002).

Both Eqs. (2) and (3) correspond to a "cookie-cutter" strategy, widely used in subsurface modeling to model strata (and facies) in reservoir grids (e.g., Johnson & Jones, 1988; Lemon & Jones, 2003; Pyrcz & Deutsch, 2014b; Swanson, 1988). It was also used early on for velocity modeling by structurally complex domains by Gjøystdal et al. (1985). In that case, the indicator functions for the J regions are determined by boundary representation and Boolean operations. In basin modeling, Mello and Henderson (1997) proposed to create a boundary representation of the regions R_J by stitching geological interfaces. Overall, this nested modeling strategy is essential to reflect the hierarchy of heterogeneities encountered in the subsurface.

Another general concept is to use some intermediate variable(s) $\mathbf{u}(\mathbf{x})$ to represent some geometric aspects of subsurface models. In Eq. (1), this amounts to replacing the position variable $\mathbf{x}$ by one or several variables $\mathbf{u}(\mathbf{x})$, yielding

$$m(\mathbf{x}) = m(\mathbf{u}(\mathbf{x})) = \sum_{k=1}^{K} m_k \psi_k(\mathbf{u}(\mathbf{x})), \tag{7}$$

where $\mathbf{u}(\mathbf{x})$ can be defined by another linear combination of basis functions χ_i:

$$\mathbf{u}(\mathbf{x}) = \sum_{i=1}^{I} u_i \chi_i(\mathbf{x}). \tag{8}$$

As in Russian dolls, these geometric basis functions $\chi_i(\mathbf{x})$ may also be truncated or designed to be conformable to some region boundaries.

The velocity values in Fig. 6B illustrate this approach, as the corner-point grid bears an indirect representation of the $\mathbf{u}(\mathbf{x})$ coordinates by the regular indexing of the grid blocks. This indexing was used to extrapolate the velocity values from sparse observations so that directions of velocity anisotropy followed stratigraphic orientations. Fig. 6D describes a more general function $\mathbf{u}(\mathbf{x}) = [u(\mathbf{x}), v(\mathbf{x}), t(\mathbf{x})]$ representing the stratigraphic geometry. $t(\mathbf{x})$ is shown on the one-dimensional profile. In this example, relatively coarse linear basis functions are used to represent the intermediate variable $\mathbf{u}(\mathbf{x})$ on a tetrahedral mesh. The finer regular piecewise constant basis functions represent properties $m(\mathbf{u})$.

This type of approach is common in the modeling of stratigraphic domains, as they allow to align petrophysical features on the structural directions, see for example, Swanson (1988), Mallet (2004), Lomask, Guitton, Fomel, Claerbout, and Valenciano (2006), and Wu (2017).

To summarize, the examples shown in Fig. 6 illustrate the interest of combining different sets of basis functions to decouple the representation of model geometry and the representation of model properties. Geological concepts are introduced as specific basis functions, as truncation of some basis functions, or as composition of basis functions. The consequences of these model choices are important, because they impact the ability of the models to approximate the true subsurface features. As the number of parameters and the relationships between these parameters change from one parameterization to the next, these choices also have a strong impact on computational performance, and on the structure of model space in uncertainty quantification and subsurface inverse problems.

3.1.2 Stratigraphy and representations of geological time

So far, we have mainly described subsurface parameterization in a mathematical way, starting from the petrophysical parameter field of interest. We now proceed to discuss model choices by starting from essential geological concepts and how they translate into constraints or into the parameterizations presented in the previous section.

As discussed in Section 1.1, geology puts a significant focus on chronological considerations to analyze and interpret observations. This temporal (or historical) reasoning can partly be integrated into geomodeling strategies.

As shown in Fig. 1, a part of the history of rock formation and subsequent deformation phases is classically captured by a stratigraphic column. A stratigraphic column (Fig. 8) describes the vertical stacking of units, the nature and the characteristics of the rocks in each unit, and the type of stratigraphic surfaces bounding each unit. Following the superposition principle Steno (1669), the vertical axis in a stratigraphic column has both a spatial and temporal meaning. This description, valid at a certain scale, has significantly been used in geological modeling, as it organizes geological knowledge in a systematic way (e.g., Calcagno et al., 2008; De la Varga, Schaaf, & Wellmann, 2018; Perrin & Rainaud, 2013; Zhu, Zhang, Li, Pan, & Sun, 2012).

Some freely available kinematic models exist to explicitly model deposition and deformation phases, for instance Noddy (Jessell & Valenta, 1996), PyNoddy (Wellmann et al., 2016) and the online web app Visible Geology. These kinematic models are simple and computationally efficient; they can generate a large diversity of three-dimensional models, but they cannot directly honor the diversity of observations described in Section 1.3.

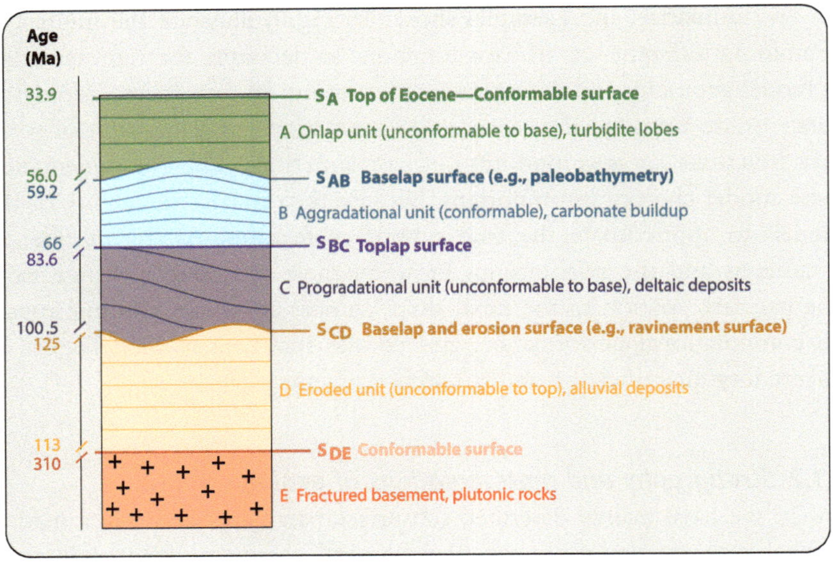

Fig. 8 The stratigraphic column: a summary of geological history recorded in the sediments. The conformity or truncation of beds within larger stratigraphic units can be related to tectonic and stratigraphic history. In this example, beds within the units correspond to chronostratigraphic surfaces.

In structural modeling, the geometry of strata is generally treated by spatial interpolation (Section 3.2), while the unconformities between strata are modeled by independent interpolations followed by truncation. For example, the basis functions for horizons or rock units are canceled above an erosion surface (Fig. 7). Conversely, the basis functions corresponding to horizons or rocks younger than a given baselap surface are canceled below that baselap surface. Geological time is, therefore, treated indirectly by selecting the type of truncation operation which reflects the stratigraphic history.

More generally, cross-cutting relationships between rock units are essential to determine the relative ages of geological objects. For instance, a layer cut by a dyke or by an evaporitic intrusion is older than the intrusive unit. Again, this can be handled by independent interpolations of the boundary surfaces of layers and intrusions and truncation of older rocks by younger ones. This reasoning, corresponding to Eq. (3) and illustrated in Fig. 7C, and has been implemented on various numerical representations (Calcagno et al., 2008; De la Varga et al., 2018; Gjøystdal et al., 1985; Johnson & Jones, 1988; Lemon & Jones, 2003; Swanson, 1988; Zeng, Backus, Barrow, & Tyler, 1998; Zhu et al., 2012). Geological surfaces can also be created or edited so as to directly enforce the termination of surfaces on other surfaces following the prescribed relationships (Brandel et al., 2005; Caumon, Sword, & Mallet, 2003; Mallet, 1992; Mello & Henderson, 1997; Perrin, Zhu, Rainaud, & Schneider, 2005), which corresponds to implementations of Eq. (2) illustrated in Fig. 7B.

Another way to treat time in structural modeling is by explicitly or implicitly representing the geological time $t(\mathbf{x})$ as a function of space (Eq. 7). A classical way of doing so is to include layering information within stratigraphic columns. Layering essentially informs the interpolation between the bounding surfaces of each stratigraphic unit. Three main interpolation functions have been defined, depending on the type of horizons bounding the units of a stratigraphic column (Johnson & Jones, 1988; Mallet, 2002; Perrin & Rainaud, 2013; Swanson, 1988; Zeng et al., 1998): proportional between top and base (conformable: unit B in Fig. 8), eroded (conformable to base: unit D in Fig. 8), baselap (conformable to top, Fig. 8). A fourth style combining eroded and baselap also exists (unit D in Fig. 8) and can be modeled by splitting the considered unit by a conformable horizon. In simple cases such as described in Fig. 2A, geological time $t(\mathbf{x})$ can be determined by simple linear operations advecting the key stratigraphic surfaces within the adjacent units (see further details and references in Section 3.2.1). These geometric styles are a simplification of more advanced genetic concepts (see for instance Catuneanu et al., 2011). A better integration of concepts is the topic of recent research. In particular, due to the difficulty of determining the exact age of sediments, uncertainty about stratigraphic history may be described using several possible stratigraphic columns built from the data at hand and from various depositional concepts (Borgomano, Fournier, Viseur, & Rijkels, 2008; Edwards, Lallier, Caumon, & Carpentier, 2018; Lallier et al., 2012).

3.1.3 Tectonic and epigenic structures

The Earth crust is affected by deformations and circulation of fluids under high temperature and pressure. These processes generate tectonic structures (which correspond to various types of rock deformation) and epigenic processes (which concern the transformations of rocks due to fluid circulations).

Tectonic events can also be formalized in terms of intrusions, fold and fault characteristics. In simple terms, intrusions are due to rocks behaving like viscous fluids (magma and evaporitic rocks at geological time scales), folds accomodate elasto-visco-plastic rock deformation, whereas faults correspond to brittle deformation localized on a slip surface. The transition between these modes of deformation is not simple to establish, and may vary depending on the considered temporal or spatial scale of interest.

Intrusions can have very complex shapes and topologies. Indeed, fluid-like rocks may follow intricate paths in the subsurface, see for instance Hudec and Jackson (2007). As intrusions cross-cuts surrounding rocks, they are considered younger, event though they may be coeval with the deposition of some surrounding rocks (e.g., Giles & Lawton, 2002). The geometry of intrusions is generally determined by geometric interpolation. Relations with surrounding structures are treated by surface stitching or truncation using the same operations as with baselap stratigraphic surfaces.

Folds are typically encountered in compressive settings to accommodate shortening. In this case, folds are characterized by a higher curvature in the shortening direction than in the orthogonal direction. Folds may also exist in extensive or strike-slip regimes, where they often appear ahead of faults. In general, folded horizons should be globally developable (i.e., may be flattened with minimal distortion). This type of principle can be used to check the likelihood of a particular model (e.g., Gratier & Guillier, 1993). In the specific case where horizon data consist of contour lines, Thibert, Gratier, and Morvan (2005) also proposed a direct method that directly creates horizon contours lines.

A fault is generally considered as a zero-thickness slip surface that juxtaposes rocks of different nature due to the accumulation of displacement. Faults often have complex growth histories. Stress can be relaxed by frictional slip on the fault surface, by propagation of the fault tip, and by coalescence with other faults (Marchal, Guiraud, Rives, & van den Driessche, 1998). The deciphering of tectonic structures associated to faults involve several kinematic models which describe how deformation localizes around a fault in various contexts (e.g., Cardozo & Aanonsen, 2009).

In our Pyrenean outcrop example, a fault-propagation mechanism is invoked by Alonso and Teixell (1992) to explain present-day observations: the subhorizontal thrust faults separating the thrust sheets root into a deep detachment level below the Marboré Formation. Each fault developed toward the south, and the corresponding shortening was accommodated by a fold ahead of the fault tip. The fault has then accumulated displacement and has grown to cut the folded structure. Although it is not directly visible, a thin level of evaporite probably lubricated the decollement and filled the gaps by viscous flow between more competent rocks.

Faults are characterized by their orientation, size, roughness, and displacement. Some of these parameters can be observed locally and are also often described in statistical terms (e.g., Kim & Sanderson, 2005). Relationships between geometric parameters and causality with hydromechanical processes are complex because of mechanical and hydrodynamic heterogeneities. In particular, interactions between faults are complex, leading to complex topological and geometrical configurations. Indeed, a fault may splay at the tips (Marchal et al., 1998) or may grow until it branches onto an older fault. If propagation continues, the older fault is displaced by the younger one. When two conjugate branching faults intersect, they can accumulate displacement alternatively, creating X-shaped branch lines (Ferrill, Morris, Stamatakos, & Sims, 2000). Fault may also be eroded by stratigraphic surfaces.

In structural modeling, faults are generally treated as three-dimensional surfaces. Determining the geometry of fault surfaces calls for interpolating between observations. The amount of fault displacement is determined by interpolation from observation data (e.g., Calcagno et al., 2008; Mallet, 1992), by interactive techniques (Caumon et al., 2003) or using specific near-field fault models (Calcagno et al., 2008; Georgsen, Røe, Syversveen, & Lia, 2012; Godefroy, Caumon, Ford, Laurent & Jackson, 2018; Laurent, Caumon, Bouziat, & Jessell, 2013). As compared to other geological surfaces which always terminate onto other surfaces, fault surfaces can have tip lines not associated to any other surface. This can be translated into fundamental topological rules for geomodeling systems (Caumon et al., 2003). This also raises practical topological and geometrical challenges to appropriately represent discontinuous properties across these internal discontinuities, see, for example, Chapter 2 of Mallet (2002), Moës et al. (2002), and Cherpeau, Caumon, and Lévy (2010).

Rocks may be transformed after their formation, e.g., by dissolution, precipitation and other mineral transformations in the subsurface. The features in the resulting epigenic objects are controlled by thermo-hydro-chemical processes which may span very long periods of time. Spatial prediction of these features is generally very difficult. Therefore, epigenic objects are often modeled as an overprinting of the initial structural models, using truncation as in Eq. (3), often combined with a latent variable as in Eq. (7) as in pluri-Gaussian methods (Armstrong et al., 2011; Galli, Beucher, Le Loc'h, & Doligez, 1994) and some orebody delineation methods (e.g., Martin & Boisvert, 2017). Depending on the geological context, it can be interesting to identify the geological meaning of this latent variable, e.g., the distance to fractures or other geological interfaces (Henrion, Caumon, & Cherpeau, 2010; Rongier, Collon-Drouaillet, & Filipponi, 2014).

3.1.4 Modeling workflows

In Chapter 6 of Perrin and Rainaud (2013), it is recommended in general to create and assemble geological surfaces by increasing geological age. This is justified by the principle that "an older geological event cannot modify a younger one". This principle is sound and used in kinematic modeling methods such as Noddy (Jessell & Valenta, 1996; Wellmann et al., 2016). It also led to an interesting geomodeling approach which assembles geological boundaries by topological operations completely determined by chronological considerations (Brandel et al., 2005; Perrin et al., 2005). However, this type of approach raises challenges because relative chronologies between objects can be difficult to establish. Moreover, all available observation points characterize the present geometry and topology of the system and not the past state. this may explain why the most common approach in commercial geomodeling packages (e.g., SKUA-Gocad, Petrel, and Geomodeller) is to start by first modeling geological discontinuities such as faults and intrusions, and then to focus on stratigraphic surfaces.

Other model building orders can also be deemed convenient depending on the data at hand. For instance, one may want to start with interfaces which are most sampled or most visible from available data. Starting by modeling objects which are deemed to have most impact on the modeling outcome may also be an option (Ringrose & Bentley, 2015).

For better or for worse, it is therefore not always possible to relate the modeling order of objects to geological chronology. Consider, for example, a salt intrusion truncating the surrounding strata: one may interpolate both geometries independently, and then cut the sediments; alternatively, one may first model the geometry of the intrusion, then interpolate the geometry of the strata to end on the unconformity (see Fig. 7). These two strategies can probably reach the same final result, but they certainly have an impact in terms of how easy model updating can be achieved and on overall model building performance and memory cost.

In today's practice, this arbitrary modeling order often implies iterations, and model updating as construction proceeds (Caumon, Laurent, et al., 2013). These iterations translate how uncertainty is iteratively reduced by letting experts incorporate their geological knowledge interactively to obtain the "best" possible model. Approaches which aim at sampling uncertainties (Section 2) rely on mathematical rules such as the truncations discussed above. In both cases, interpolation between observations points is an essential component to come up with realistic geometry of the considered geological surfaces.

3.2 Numerical representations and interpolation of geological structures

This section further delves into the possible ways to numerically describe and build structural geometry from available data and described some possible choices and meanings for the basis functions $\chi_i(\mathbf{x})$ representing the model geometry.

Determining the geometry of geological structures from sparse data has long been a concern for geologists. Numerical methods focus either on:

- Directly building geological interfaces (explicit structural modeling methods). In this case, each geological interface is defined by interpolating between the nodes of a two-dimensional orientable graph. As surfaces are embedded in three-dimensional space, surfaces do not necessarily define a closed volume, and modeling involves projections of the data points onto the surface.
- Building a scalar function whose equipotential surfaces are the geological boundaries of interest. The scalar function may be seen for instance as a signed distance to the interface or as a relative geological time function. This type of method is, in general better posed, as no data projections are involved.

3.2.1 Explicit methods: Maps and extrusion approaches

The advent of computers in the second half of the 20th century led to so-called "automatic contouring" techniques, where the goal was to identify the elevation (or depth) $z(x, y)$ of a particular geological surface at the geographic location (x, y) from a set of I sample points $(x_i, y_i, z_i)_{i=\{1, \ldots, I\}}$ (Hardy, 1971; Olea, 1974; Walters, 1969). These methods can only address relatively simple geological configurations as described in Fig. 2A. However, they have been much used out of consideration for their simplicity and computational efficiency. This section summarizes the main variants of map-based approaches and introduces some mathematical interpolation principles which are common to all geomodeling methods.

Early on, the various formulations of the problem were either very practical (asking how to program a contour plotting device from the data) or more theoretical (finding a semianalytical solution by combining a series of polynomial or Fourier basis functions). The latter case can be posed as a classical interpolation problem, see, for example, Hardy (1971), where the aim is to estimate the coefficients c_i of

$$z(x, y) = \sum_{i=1}^{I} c_i \chi_i(x, y), \tag{9}$$

where $\chi_k(x, y)$ is some basis function centered at location (x_k, y_k), for instance $\chi_i(x, y) = [(x-x_i)^2 + (y-y_i)^2]$.

In the presence of I data points $(x_i, y_i, z_i)_{i=1, \ldots, I}$, identifying $z(x_i, y_i)$ with z_i forms a linear system whose resolution provides the coefficients c_i:

$$[\chi_j(x_i, y_i)] \cdot [c_j] = [c_i]. \tag{10}$$

As discussed by Hardy (1971), the use of additional data such as slope information could be addressed by a least-squares system using the analytical derivatives of $\chi_i(x, y)$ and/or by adding polynomial terms $p_l(x, y)$ to (9):

$$z(x, y) = \sum_{i=1}^{I} c_i \chi_i(x, y) + \sum_{l=1}^{L} d_l p_l(x, y). \tag{11}$$

Kriging is also significantly used in contouring (e.g., Dubrule, 1984; Lemon & Jones, 2003; Olea, 1974). A complete coverage of kriging is out of the scope of this paper, but we refer to Cressie (1990) for a historical perspective and to Chiles and Delfiner (2012) for a thorough description. Essentially, kriging is preceded by a spatial data analysis phase, which aims at statistically characterizing the variable of interest (here the depth $z(x, y)$), considered as a spatial random field. In this frame, each data point z_i at location (x_i, y_i) is seen as the local outcome of a spatial random process. In the universal kriging formalism, the depth at location x, y is given by a linear combination of the data values plus some polynomial trend function:

$$z(x, y) = \sum_{i=1}^{I} \lambda_i(x, y) z_i + \sum_{l=1}^{L} d_l p_l(x, y). \tag{12}$$

The weights of this combination are obtained by minimizing the estimation variance of z (i.e., the spread between the model z and the unknown true depth $\tilde{z}$), under unbiasedness condition (i.e., stating that the average error between the model and the truth should be centered on zero). To determine the kriging weights λ_i, geostatistics uses a variogram model, which captures important features about spatial variability, assuming it only depends on distances. In the end, the weights are obtained by solving a linear system fully determined by the variogram model and by the geometry of the data. When all data points are used for the kriging estimation, some algebraic manipulations produce the dual kriging system of Eq. (12), which has exactly the same form as Eq. (11), χ_i corresponding to the covariance model. One practical benefit of the geostatistical approach is in the process to determine the shape of the basis functions ϕ, which are inferred from the available data using variogram analysis (Dubrule, 1984).

Other interpolation methods such as inverse distance interpolation and natural neighbor interpolation have also been used in geological modeling (Lemon & Jones, 2003; Ming, Pan, Qu, & Ge, 2010). Like kriging, both methods are linear estimators:

$$z(x,y) = \sum_{i=1}^{I} \lambda_i(x,y) z_i.$$

Inverse distance weight each data point as using its relative inverse distance to the location to be estimated. Natural neighbors (Sibson, 1981) involve Voronoi diagrams compute the weights: λ_i is given by the relative area of overlap the domain nearest to (x, y) than to all the data points and the domain nearest to data point i than to any other data point. This robust weighting scheme amounts to locally adapting the support of the underlying basis functions χ_i depending on data configuration. This is interesting for geoscience applications (Watson, 1999), but needs specific care in extrapolation (Bobach, Farin, Hansford, & Umlauf, 2009). More generally, both methods are easier to apply than kriging, as less parameters and no statistical inference are involved. Conversely, the geostatistical approach provides a procedure to control the spatial structure of the solution and to account for other data types, such as slopes (Hardy, 1971).

Many important details matter in these map-based approaches, from the well-posedness of the system to the management of faults, which are generally treated as discontinuities (Mallet, 1984; Marechal, 1984; Pouzet, 1980). Also, once maps have been interpolated in three dimensions, information about unconformities can be used to truncate the various surfaces and obtain the final layer geometry. For all these operations, implementation vary depending on the chosen numerical representation: rectangular grids (Fremming, 2002; Mallet, 1984; Marechal, 1984; Pouzet, 1980; Swanson, 1988) or triangulated surfaces (Kessler, Mathers, & Sobisch, 2009; Lemon & Jones, 2003; Zhu et al., 2012).

Another aspect relates to the inability of map-based methods to represent and visualize structures having several elevations for the same geographic coordinates, for instance ore bodies, overturned folds, or horizons affected by reverse faults.

When faults intersect in map view but not in vertical view, this limitation can be addressed by defining a direction field tangent to the fault surfaces, see

(Fremming, 2002) and Mallet (2002, Chapter 8). This approach, often called pillar model, is illustrated in Fig. 9B, where pillars are shown as red lines. The horizon surfaces are then modeled by interpolating their *apparent* depth (or layer thickness) along these pillars, by treating faults as discontinuities. All the map-based methods are then applicable by replacing the vertical direction by the pillar direction.

As visible when comparing Fig. 9A and B, this method is limited to representing fault that have approximately the same dips. An alternative (or complementary) approach consists in extrapolating faults so as to completely subdivide the domain of interest into closed fault blocks, then building strata within a larger domain and trimming by the resulting fault blocks (e.g., Hoffman & Neave, 2007a).

As discussed by Fremming (2002) and Mallet (2002, Chapter 8) the above pillar-based approaches (and the map-based methods (Lemon & Jones, 2003)), are often used in a hierarchical way to adapt to hierarchy of stratigraphic series. A few key horizons are first created (e.g., from reflection seismic interpretations), then intermediate horizons are interpolated between these key horizons using available data (e.g., borehole data). In the context of stratigraphic uncertainty modeling, Abrahamsen (1993) also used this principle to propose a geostatistical framework for the coherent conversion of a series of stratigraphic surfaces from the reflection seismic time domain to the depth domain. These methods are often terms extrusion methods or 2.5D methods, as they use two-dimensional algorithms to create three-dimensional surfaces. As horizons are all represented as depths values along each pillars, minimal and maximal layer thickness can be enforced during interpolation. Stratigraphic unconformities can be obtained by setting consecutive horizons to the same depth value.

Another benefit of these approaches is that they relatively easily allow to create volumetric grids supporting petrophysical and physical modeling (Fremming, 2002; Hoffman & Neave, 2007a; Johnson & Jones, 1988; Swanson, 1988). This explains why map-based and pillar-based models, in spite of their representational limitations, are vastly used in reservoir and near-surface modeling applications. Various implementation variants exist in software such as Petrel, Gocad, RMS, or Jewel Suite.

3.2.2 Full 3-D modeling approaches: Explicit

Instead of describing the depth as a function of the geographic position, a three-dimensional surface mesh can be used to describe geological

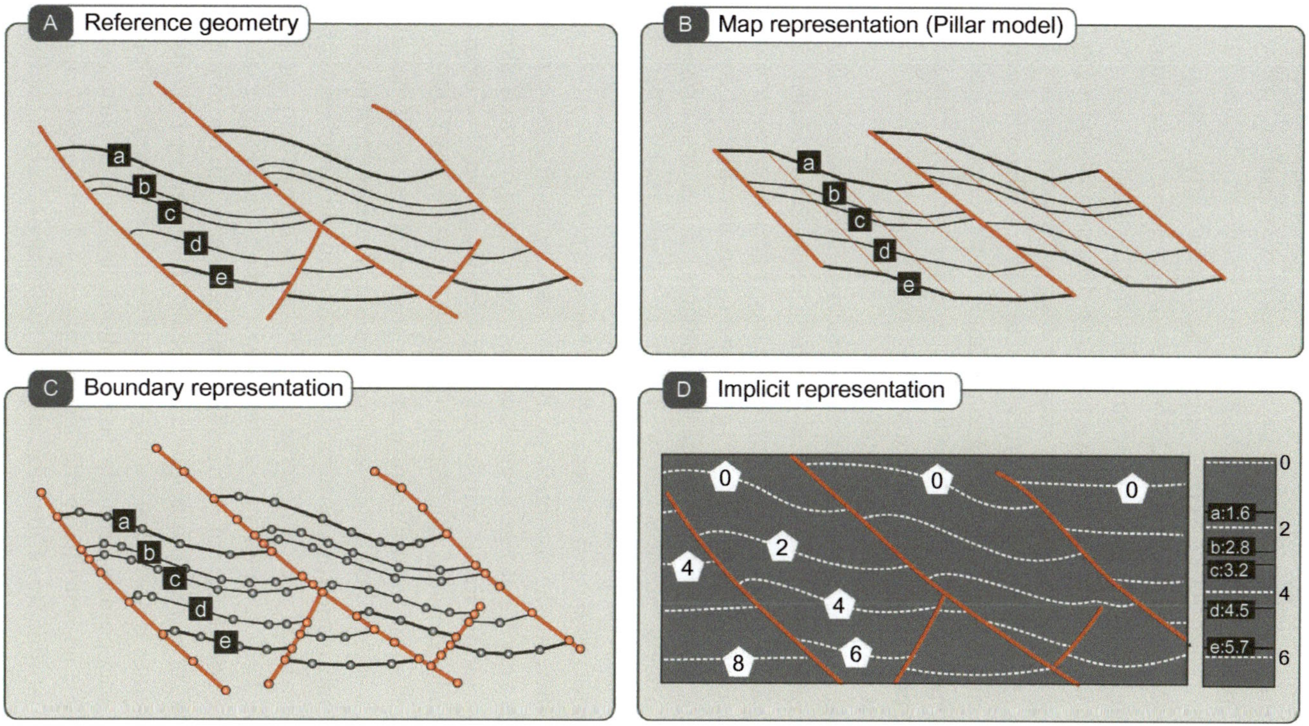

Fig. 9 Cross sections of various numerical representations of geological structures. *Red lines* correspond to fault surfaces and *black lines* correspond to five stratigraphic surfaces (a, c, b, c, d, and e). (A) Reference continuous model. (B) Pillar-based or extrusion methods (Section 3.2.1) use two-dimensional depth or thickness maps computed along a direction field tangent to the main faults. (C) Boundary representations (Section 3.2.2) using piecewise linear surfaces with three-dimensional coordinates to define sealed geological volumes. (D) Implicit representations (Section 3.2.3) treat some surfaces—here, stratigraphic horizons—as iso-values of a three-dimensional scalar field (figured with the *white dashed lines* and the corresponding values).

interfaces. Surface meshes consist mainly or triangulated surfaces (also called triangular irregular networks or TINs) (Mallet, 1992; Ming et al., 2010) or parametric surfaces such as NURBS (Auerbach & Schaeben, 1990; de Kemp & Sprague, 2003; De Paor, 1996; Fisher & Wales, 1992; Sprague & de Kemp, 2005). As these approaches have already been reviewed by Sides (1997) and Caumon et al. (2009), we only give here a brief summary or the methods so as to discuss their strengths and limitations.

Triangulated surfaces consist of a set of three-dimensional nodes $\{\mathbf{x}_1, \ldots, \mathbf{x}_K\}$ connected by triangles. In the simplest case, each node bears a linear basis function χ_k equal to 1 at the vertex k and to zero at all the other vertices. The continuum of points belonging to a triangle is, therefore, described by linear combination of the three basis functions centered on the triangle vertices. This type of mesh can approximate the shape of any free-form surface provided that the mesh density is sufficient.

A first approach to create triangulated surfaces consists in associating curves defined on parallel cross sections (Caumon et al., 2009; Mallet, 2002; Sides, 1997). These methods are applicable to cylindrical or conical structures, but need many cross sections to address domains affected by nonparallel faults, as the topology of faulted surfaces change from one cross section to the next. Surfaces can be also created between boundary lines and borehole data, for instance by Delaunay triangulation and mesh refinement, see Sides (1997) and Mallet (2002, Chapter 3).

As all of these methods connect points and lines by linear segments, they tend to create very angular geological surfaces. Smoothing methods allow to obtain more realistic shapes while honoring subsurface data. For example, discrete smooth interpolation (DSI) method (Mallet, 1989, 1997, 2002) uses numerical optimization to minimize the surface roughness while honoring available data. The principle is to start from an initial surface, then to express all pieces of information as a linear relation applied on the coordinates of the surface nodes. Finally, each of these linear relations is solved in the least-squares sense. For example, the roughness criterion can be locally expressed by setting the coordinates of each node to the average of the coordinates of its neighbors (note that this requires a specific treatment of boundaries). Another criterion is defined to locally move the surface toward data points: A particular data point $\mathbf{x}_D$ that the surface should honor must first be projected onto the nearest surface triangle ($\mathbf{x}_A$, $\mathbf{x}_B$, $\mathbf{x}_C$). Denoting (a, b, c) the barycentric coordinates of the projection in the triangle, on can write that the triangle includes the data point $\mathbf{x}_D$ if

$$\mathbf{x}_D = a \cdot \mathbf{x}_A + b \cdot \mathbf{x}_C + c \cdot \mathbf{x}_C. \tag{13}$$

More simply, a particular surface node may be locked so that is position is not altered during interpolation, which make it possible to exactly honor some data points in they are included in the surface mesh. Mallet (2002) presents several other linear criteria or inequality constraints that can be used to honor other types of information, for instance local surface orientation, contacts between surfaces or thicknesses. In the end, all these linear equations are weighted and assembled in a least-squares system to find the surface shape.[g] The roughness is essential to ensure the well-posedness of the DSI system, as there are in general fewer data points than surface nodes. This method, together with the aforementioned surface creation and cutting algorithms, forms the core of the Gocad geomodeling platform developed in the 1990s (Mallet, 1992).

These methods to create and smooth triangulated surfaces are well-defined mathematically and can be applied in an automatic way to create the various surfaces of interest independently one from another. However, when the available data cannot be projected onto a plan without changing the neighborhood (e.g., in the case of an overturned fold as in Fig. 2B or of a complex intrusion as in Fig. 2C), the creation of the initial surfaces generally involves manual operations to reach the desired result. One then needs to decompose the data to create and then merge subplanar patches or to locally change of the directions of attraction between the surface and the points (Caumon et al., 2009).

Instead of linear basis functions, NURBS and Bézier parametric surfaces essentially use higher order basis functions interpolating a control network of surface nodes to describe the geometry of surfaces. This makes it possible to represent surface geometry with a smaller number of nodes than with piecewise linear surfaces. This also allows to design complex folds and overturned surfaces by interactively moving a few control nodes in space. de Kemp and Sprague (2003) and Sprague and de Kemp (2005) use this feature for instance to interactively design complex fold geometries by combining orientation data and graphical interactions. In general, these parametric surfaces use a tensor product of parametric curves. As a result, surfaces essentially have a rectangular shape and adaptive refinement is more difficult than with triangulated surfaces. As show by Zhang et al. (2018), T-splines open interesting perspectives to address these challenges and manage complex interactions (Fig. 2) in the design of parametric geological surfaces.

[g] As the form of the system depends on the initial surface, the least-squares DSI system is solved iteratively by updating the coefficients of Eq. (13) (and of the other data terms) between each iteration.

Indeed, whether triangulated or parametric, the relations between surfaces created independently are generally inconsistent with the geological principles reviewed in Section 3.1. In particular, surfaces created independently tend to cross one another or to leave gaps within the domain, which violates essential geomodel validity conditions (Caumon, Lepage, Sword, & Mallet, 2004).

To avoid this, a first option is to carefully design the surface construction methods to avoid inconsistencies in the first place. Consider for instance surfaces created by associating parallel cross sections. In this case, specific sections need to added by the modeler to delimit fault tip lines and where ore bodies terminate (Sides, 1997). The branch lines between surfaces are then linearly approximated between parallel sections. Another approach is to compute the fault surfaces and the corresponding branch lines. Horizons can then be created by picking the horizon cutoff lines on the fault surfaces used as cross sections. Each horizon can then be created automatically by triangulating the interior of the so-called fault polygons using the available data points.

As these approaches are tedious, another option is first to create explicit surfaces covering the whole domain from all the data at hand, then corrects the problems using automatic or semiautomatic geometrical and topological operations: cutting surfaces (e.g., horizons by faults), surface trimming (e.g., faults by fault tip line of faults by other faults), extrapolation (e.g., to enforce the branching of two neighboring faults), and constrained interpolation to update surface geometry while maintaining the topological relationships (see review of Caumon et al., 2009). These methods borrow a lot to computer-aided design (CAD) and may have been applied early on by editing the mesh of triangulated surfaces (Mallet, 1992) and on parametric surfaces by truncating the parametric surfaces (Gjøystdal et al., 1985).[h] In the end, mending all surfaces along their common boundary generates a boundary representation (Fig. 9C).

All the above methods imply a significant level of user interaction and validation, which require practice and skills from the modeler and involve subjective choices. Several authors have proposed avenues to automate (or at least facilitate) geomodel building and make it both faster and more reproducible. This includes for instance automatically removing horizon data in

[h] Again, changing the mesh of a triangulated horizon surface to align on a fault surface amounts to make its basis functions conformal to the fault (Eq. 2), whereas truncating a parametric horizon surface depending on the side of the fault correspond to Eq. (3).

the neighborhood of faults to reduce the risk that a horizon point attracts the surface on the other side of the fault. These practically important heuristics are implemented in commercial software but are seldom described in scientific papers.

Notably, another strategy to increase automation uses a completely different approach based on Voronoi diagrams. These methods are rooted in computational geometry approaches, which are often used for general-purpose surface reconstruction from point sets (Amenta & Bern, 1999). In geoscience applications, the main idea is to reconstruct all the surfaces at once by aligning the faces of Voronoi polyhedra on geological interfaces. This is achieved, either by a careful processing of data points (Courrioux et al., 2001), Delaunay triangulation of all data points followed by incremental corrections (Beni, Mostafavi, & Pouliot, 2009) or by numerical optimization (Hale & Emanuel, 2003; Merland, Caumon, Levy, & Collon-Drouaillet, 2014). These methods automatically provide a reasonable approximation of geological interfaces but are sensitive to data density and quality. They actually represent a first class of methods which consider volume data structures to create geological surfaces.

3.2.3 Full 3-D modeling approaches: Implicit

Implicit modeling techniques represent geological surfaces as iso-values (or level sets, or equipotentials) of a three-dimensional scalar field $s(\mathbf{x})$. This idea is quite old in itself (Houlding, 1994; Lajaunie et al., 1997; Mallet, 1988), but it only became practical during the 2000s as computer memory increased.

Considering a single geological surface (ore body, horizon, fault), the idea of implicit methods is to compute the scalar field $s(\mathbf{x})$ so that all data points sampling the surface have the same scalar value. In this case, it is convenient to see this scalar field as the signed distance function to the surface, being negative when $\mathbf{x}$ lies below (or inside) the surface and positive when $\mathbf{x}$ lies above (or outside) the surface, and null on the surface.

The essential principles to interpolate the scalar field $s(\mathbf{x})$ between data points are the same as with map-based interpolation (Section 3.2.1). However, stating that $s(\mathbf{x})$ should be equal to zero at all data points $\mathbf{x}_i$ is not sufficient to form a well-posed system. Indeed, a trivial solution would be that $s(\mathbf{x}) = 0$ everywhere. Additional conditions are therefore needed so that the gradient of $s(\mathbf{x})$ is never null. Adding extra artificial data points with nonzero values, adding some orientation measurement or forcing the gradient norm to be unitary is therefore needed to come up with a viable solution. In the case of ore body or salt modeling, data consist of internal and external points, so these can be handled by inequality constraints during least-squares minimization (e.g., Frank et al., 2007) or choosing appropriate positive or negative values for the points away from the interface (Martin & Boisvert, 2017). In the case where all data points are on the surface being modeled, at least one orientation point is needed (Calcagno et al., 2008; Caumon, Gray, et al., 2013), or, equivalently, the points need to be duplicated and offset before interpolation given some initial normal estimation (Carr et al., 2001).

Two main numerical methods have been proposed to create individual implicit surfaces from sparse data points:

- As in Eq. (11), meshless methods estimate the scalar field $s(\mathbf{x})$ as a linear combination of I basis functions χ_i and monoms $p_l(\mathbf{x})$:

$$s(\mathbf{x}) = \sum_{i=1}^{I} c_i \chi_i(\mathbf{x}) + \sum_{l=1}^{L} d_l p_l(\mathbf{x}) \tag{14}$$

This general form includes radial basis function (RBF) interpolation (Carr et al., 2001; Cowan, Beatson, Fright, McLennan, & Mitchell, 2002; Hillier et al., 2014) and dual kriging (Aug et al., 2005; Calcagno et al., 2008; Chilès, Aug, Guillen, & Lees, 2004; De la Varga et al., 2018; Lajaunie et al., 1997), which have, as in the two-dimensional case, a similar system structure. In both cases, terms can be added to Eq. (14) to also use orientation data, which correspond to gradient of $s(\mathbf{x})$ (Chilès et al., 2004; Lajaunie et al., 1997). In general, the basis functions $\chi_i(\mathbf{x})$ depend on the isotropic distance $||\mathbf{x} - \mathbf{x}_i||$ between the position to evaluate $\mathbf{x}$ and the data point $\mathbf{x}_i$. As this may generate blobby geometries away from the data, Martin and Boisvert (2017) propose to use locally anisotropic distances instead or isotropic distances. In their approach the local directions of anisotropy directions are computed iteratively by solving several interpolations.

These meshless methods are very convenient but they involve solving a large and dense linear system of equations of size $(I+L)^2$. Optimization methods have been proposed (e.g., the fast multipole approach originally introduced by Greengard & Rokhlin, 1987). A recent alternative has also been proposed by Renaudeau, Maerten, Malvesin, and Caumon (2018) using an intermediate set of points which bear local moving least-squares basis functions. The overall system becomes sparse, as the basis functions have a local support, but the continuity of the solution is maintained.

- Mesh-based methods have also been proposed to compute the scalar field $s(\mathbf{x})$ (Caumon, Gray, et al., 2013; Frank et al., 2007; Laurent, 2016; Moyen, Mallet, Frank, Leflon, & Royer, 2004; Souche Lepage, & Iskenova, 2013). In this case, the domain is covered by a linear tetrahedral mesh conformable to discontinuities. The interpolation method uses numerical optimization combining linear data terms as Eq. (13) with a roughness term stating that the scalar field should vary smoothly. As in explicit methods, weights can be used to globally or locally change the relative contribution of the data and regularization terms. This can be used for instance to change from smooth fold hinges to kink-style hinges when axial surfaces are mapped in the domain (Caumon, Gray, et al., 2013).

A significant feature in implicit methods is that *several* conformable surfaces may be modeled by the same scalar field. This principle was first introduced for foliation fields (Lajaunie et al., 1997) and then widely adopted for modeling of stratigraphic series, as they typically consist of subparallel surfaces. In this case, the scalar field can be seen as a relative geological time function (denoted as $t(\mathbf{x})$ in Section 3.1). This approach is illustrated in Fig. 9D. Different ways to compute this function use the aforementioned numerical methods. As compared to the reconstruction of a single surface, a practical question is to decide about appropriate values for each horizons. Indeed, a wrong choice of values may generate local minima in the interpolated solution, which imply closed bedding surfaces that cannot be generated by stratigraphic processes. Heuristic approaches based on thickness considerations have been proposed to address this problem. (Caumon, Gray, et al., 2013; Collon et al., 2016; Collon-Drouaillet et al., 2015). Alternatively, in the case where enough data are available for the problem to be well-posed, the increment of the scalar field can be estimated as differences to some reference value (Calcagno et al., 2008; Chilès et al., 2004; De la Varga et al., 2018; Lajaunie et al., 1997; Renaudeau et al., 2018).

In the presence of unconformities, the domain is split into several conformable sequences, then Boolean operations are applied to honor the information present in the stratigraphic column (Calcagno et al., 2008; Caumon, Gray, et al., 2013; Souche et al., 2013). The treatment of faults varies between the meshless and mesh-based techniques. Both methods need to determine fault geometry explicitly or implicitly before computing the scalar field corresponding to faulted formations. Mesh-based methods may then generate a computational mesh conformal to the fault network, where faults are treated as boundaries. Both mesh-based methods and implicit methods may also create a scalar field for each fault block, followed by truncation by the fault surfaces. In the case of meshless methods, a notable option (Calcagno et al., 2008; Chilès et al., 2004; De la Varga et al., 2018) is to treat faults by adding two discontinuous drift terms to Eq. (14), each corresponding to the average offset induced by the fault. The basis function of each term is maximal at the fault center and decreases to zero at the fault tip and orthogonally away from the fault; it is activated only on one side of the fault set to zero on the other side.

Although we have focused here on the construction of implicit surfaces from sparse data points, we should also mention several approaches to create similar implicit volumes directly from reflection seismic images (Lomask et al., 2006; Wu, 2017; Wu & Hale, 2015).

In summary implicit modeling has opened new ways to automatically construct geological models with a high degree of complexity (Fig. 2), as typical in multiply deformed terranes (e.g., Jessell et al., 2014) or around dome structures (e.g., Wellmann et al., 2010a). This automation has led to significant progresses, as will be further discussed in Section 4.

3.2.4 Validating and updating geomodels

Overall, model quality is often addressed by computing the mismatch between interpretation points and the model surfaces. A complementary way is to discard some data points and use them as cross-validation data, but this approach is only applicable in the presence of dense data sets relatively to the geological complexity. Realism is somewhat more subjective to characterize. In addition to elementary coherency rules, it may be assessed visually by model scrutiny. Quantitative methods involve thickness map computations, and volumetric considerations. Restoration techniques and deformation analysis can also be used to check if the reconstructed geometry is compatible with a likely deformation history. Last but not least, computing model forecasts allows to compare them with observations (e.g., water heads or tracer data in an aquifer, arrival times, or waveforms at seismometers). In the next section, we present the various existing approaches to capture and quantify uncertainty in subsurface models.

4. Methods for uncertainty analysis in geological models

We described above the various sources and types of uncertainty in the typical workflow of geological model construction (Section 2) and the range of mathematical interpolation methods to generate single representations (Section 3). In the following section, we examine methods to analyze uncertainties in these model representations. In essence, most approaches can be summarized as an extension of a single, best-fitting deterministic model (Fig. 10A) to multiple possible models (or realizations) (Fig. 10B), for example, multiple realizations of interfaces and faults. This approach is equally well established in the field of geophysics (e.g., Tarantola, 2006) and applied (geo-)statistics (e.g., Caers, 2011; Chiles & Delfiner, 2012; Pyrcz & Deutsch, 2014a) as a way to represent and analyze uncertainties in complex nonlinear systems.

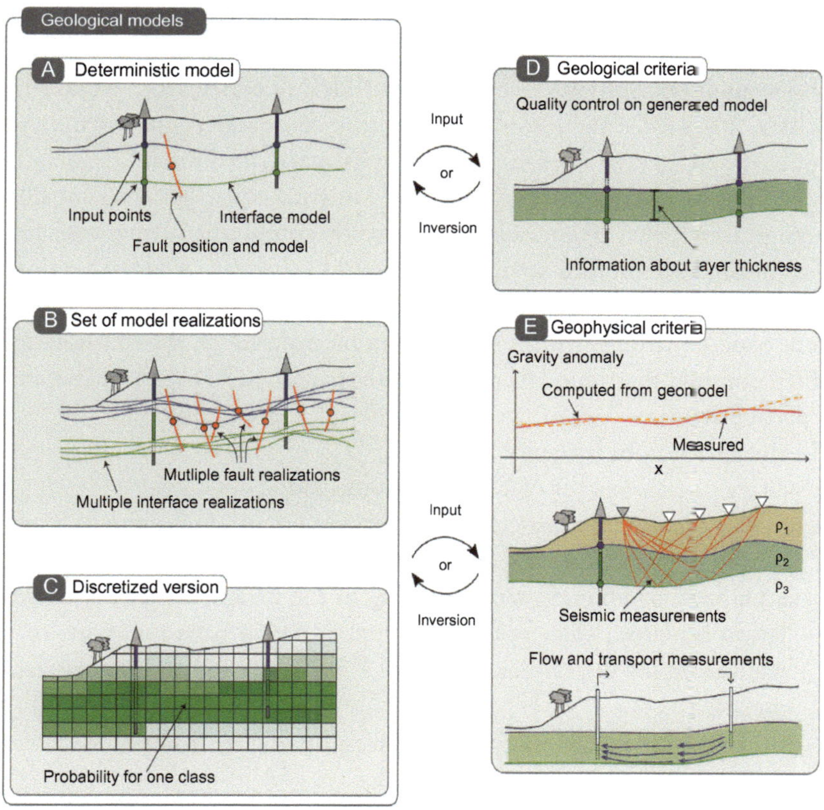

Fig. 10 The fundamental concept behind the majority of methods for uncertainty quantification in geological models is to go from one deterministic representation (A) to multiple realizations, for example, multiple interfaces and faults (B). These realizations are eventually discretized (C). Uncertainty is then reduced by confronting these geological models to ancillary geological likelihood criteria (D) or geophysical observations (E).

Generating multiple realizations involves the propagation of geological uncertainties under a particular choice about what aspect will be treated in a probabilistic way. In Section 4.1), we will review the various choices which can be made and provide concrete examples to methods which perturb the data and/ or the structural models away from the data. This generated set of models is in itself an important outcome of the analysis of uncertainties, but it raises challenges in terms of visualization and communication (Section 4.2). A second challenge is to reduce uncertainty (Section 4.3), which may be achieved by assessing the geological likelihood of the various models (Fig. 10D), or by incorporating indirect geophysical or flow measurements (Fig. 10E). Often, these two tasks involve an additional discretization and petrophysical modeling step (Fig. 10C). In this paper, we choose not to review the petrophysical modeling aspect and refer to the geostatistical resources (Chiles & Delfiner, 2012; Pyrcz & Deutsch, 2014b) and petrophysics modeling methods (e.g., Avseth, Mukerji, & Mavko, 2010). We will, however, mention some aspects of petrophysical models when reviewing examples of geophysical inversions in Section 4.3.2.

In Section 3.1, we introduced the fundamental equations describing how geological models are represented in space by combining several basis functions in Eqs. (1)–(8). However, as discussed in the remainder of Section 3, geological knowledge involves many possible variants and compositions of these basic equations. In this section, we propose more abstract notations, which consider that a model contains both spatial and nonspatial parameters, in particular:

$\mathbf{x}$: the spatial coordinates;

$\mathbf{\Phi} = \{\varphi_1, \ldots, \varphi_K\}$: a set of basis functions, and

$\mathbf{m} = \{m_1, \ldots, m_K\}$: the associated coefficient vector representing the spatial discretization of the geological model;

$\boldsymbol{\alpha}$: additional parameters corresponding to the parameters of the interpolation function, such as the exponent of radial basis functions or the parameters of the chosen covariance function, and the regularization weights;

$\boldsymbol{\kappa}$: primary geological information corresponding to data point locations and values described in Section 1.3;

$\boldsymbol{\tau}$: the topological description corresponding to relationships between geological surfaces, which represent conceptual information used to build or truncate the basis functions $\boldsymbol{\Phi}$, see Eqs. (2) and (3);

$\boldsymbol{\Omega}$: additional conceptual information or evidence not necessarily formalized in mathematical terms or not directly incorporated in the geological modeling process, for example, information about the style of deformation or the maximum burial depth of some sample points during geological history

For generality, we will denote all the above parameters as $\boldsymbol{\theta}$ in the following and clarify to which of them we refer to, if required:

$$\boldsymbol{\theta} = f(\mathbf{x}; \boldsymbol{\Phi}, \mathbf{m}, \boldsymbol{\alpha}, \boldsymbol{\kappa}, \boldsymbol{\beta}, \boldsymbol{\Omega}) \; . \tag{15}$$

In many conventional approaches, the geological uncertainties which are considered are related to the model parameters $\mathbf{m}$, while all the other aspects are kept constant (e.g., Tarantola, 2005). However, the notation for $\boldsymbol{\theta}$ clearly suggests that any uncertainty in the parameters $\boldsymbol{\Phi}$, $\mathbf{m}$, $\boldsymbol{\alpha}$, $\boldsymbol{\kappa}$, $\boldsymbol{\beta}$, or $\boldsymbol{\Omega}$ will contribute to uncertainties in the subsequent geological model. Under this framework, defining uncertainty amounts to choosing which aspects will be randomized for uncertainty quantification (UQ).

4.1 Uncertainty propagation

In practice, uncertainty quantification calls for sampling some model parameters from probability distributions. This approach, widely used in applied statistics, is commonly referred to as the *Monte Carlo approach* to forward uncertainty quantification or error propagation (e.g., MacKay, 2003). In general, Monte Carlo simulation performs independent sampling of model parameters. However, geological knowledge often suggests that parameters are related, so a concern in geomodeling applications is to reproduce the interactions between model parameters.

In the remainder of this section, we start by describing approaches which quantify uncertainties by focusing on uncertainty in input data $\boldsymbol{\kappa}$, such as interface points (Section 4.1.1 and Fig. 11A). Then, we discuss methods which also address uncertainty about model parameters $\mathbf{m}$ and interpolation parameters $\boldsymbol{\alpha}$ (Fig. 11B) for surface-based models (Section 4.1.2), explicit volume models (Section 4.1.3), and implicit models (Section 4.1.4). In Section 4.1.5, we discuss stochastic models which sample the connectivity of geological structures between observations (Fig. 11C). Finally we present ways to address conceptual uncertainties addressing the uncertainties in fundamental geological modeling rules $\boldsymbol{\tau}$ and conceptual geological knowledge $\boldsymbol{\Omega}$.

4.1.1 Sampling data uncertainty (κ)

As discussed above (Section 2.2.1), uncertainties may exist about the spatial data parameters $\boldsymbol{\kappa}$ used to build the geological model. Following the example

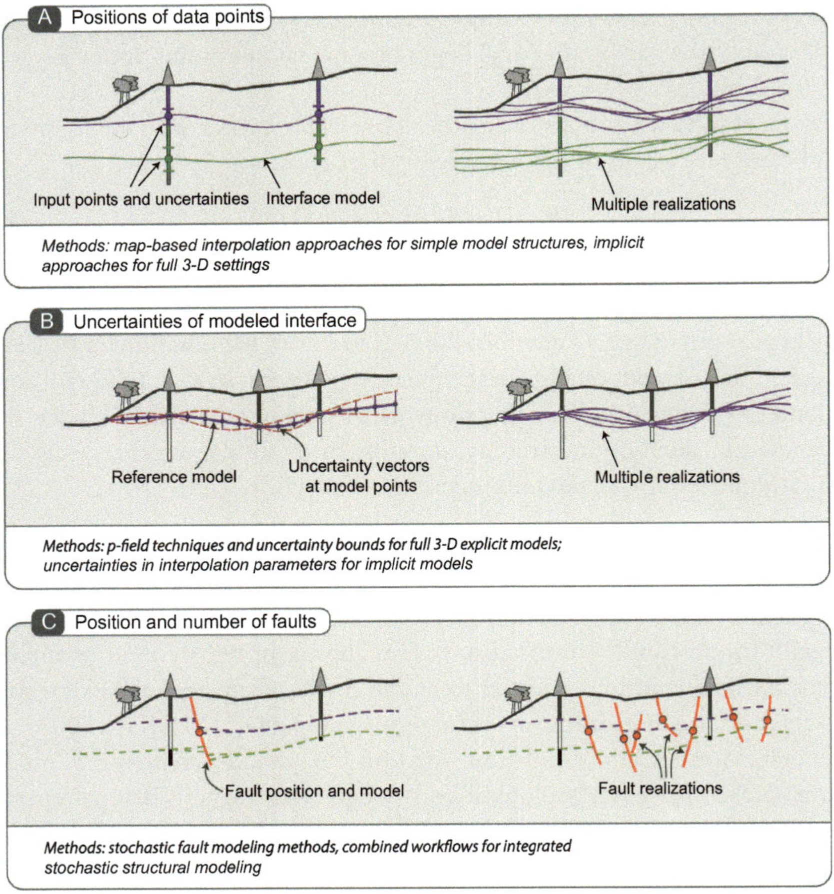

Fig. 11 Typical elements of uncertainty in generation of model realizations.

in Fig. 5D, we consider here first uncertainties about the position of a particular surface at depth. For example, Fig. 11A (left) shows vertical boreholes where the transition between layers is not perfectly observed. A reason may be that wireline logs do not allow for an exact positioning of the horizon, because core is missing, or because the transition itself is in fact gradual. As in all measurement errors, we will assume that we can describe the uncertainty about the exact horizon depth with a probability distribution.

A common approach to evaluate how these uncertainties in input parameters manifest themselves in the physical space of the model is to draw a sample from each distribution to obtain a realization of the input data set $\hat{\kappa}$. On the basis of this input data set, we can now generate the geological model $\hat{m}(\mathbf{x}; \hat{\kappa})$ using one of the interpolation methods discussed in Section 3.

As a single model realization does not provide any information about model uncertainties, a common approach is to sample n parameter sets $\hat{\kappa}^{(1)} \ldots \hat{\kappa}^{(n)}$ and to generate a geological model for each of these parameter sets (Fig. 11A, right).

This type of method has been applied both to data location and orientation measurements (Lindsay, Ailleres, Jessell, de Kemp, & Betts, 2012; Wellmann et al., 2010b; Wellmann & Regenauer-Lieb, 2012). In these cases, implicit interpolation approaches (Section 3.2.3) were used, as they allow to completely automate the model construction. In principle, all fully automatic interpolation functions can be used with this type of method. In these approaches, data values or locations are sampled independently from input probability distributions reflecting positioning or measurement errors. As discussed by Pakyuz-Charrier et al. (2018), the sampling of orientation data should consider statistical orientation models such as the Fisher–Bingham distribution, instead of an independent sampling of strike and dip. Another question with independent data sampling is whether the resulting spatial continuity of the model is realistic. Indeed, not considering spatial correlation when sampling observations may yield large layer thickness variations or unrealistic surface undulations. In some cases, parameter dependence can be determined on the basis of the measurement itself and introduced in the form of hyperparameters to the distributions. For example, uncertainties in time-depth-conversion will affect multiple seismic picks. The distributions for all of these points can then be described with respect to uncertainties in the time-depth conversion function. Similar examples are interface points determined from gravity inversion, or even multiple points on a single interface, included on the basis of expert judgment In addition, Bayesian approaches have been developed to consider additional geological criteria (Fig. 10D) in order to obtain parameter correlations (see De la Varga & Wellmann, 2016, and Section 4.3).

4.1.2 Geometric model perturbation: Surface-based methods (explicit surface parameters m and interpolation parameters α)

Instead of changing point positions and regenerating interpolations, geometric model perturbation essentially aims at adding geometric noise to an existing model (Fig. 11B). As geological structures are generally smooth, the noise is commonly correlated in space in the form of a covariance function (or variogram model). This type of method was originally introduced to assess petroleum accumulation uncertainties in oil and gas reservoirs (Delfiner & Chiles, 1977). On a map-based gridded horizon representation

(Section 3.2.1), the idea is simply to generate spatially correlated random fields for possible perturbations at each grid node. Efficient algorithms such as the turning bands or sequential Gaussian simulation may be used to generate these random fields while honoring discrete observations, see, for example, Chapter 8 of Goovaerts (1997) or Chapter 7 of Chiles and Delfiner (2012).

This simulation approach takes two sequential parameters as input: the unconditional probability distribution that the simulated random field should honor, and a model of spatial variability. Both parameters are generally assumed to be the same everywhere in space, and therefore a perturbation is generally simulated, instead of the horizon depth itself. Indeed, this separation allows the consideration of regional inclination and folding of the strata during interpolation, and then to choose appropriate statistical models to describe the residual between the interpolated model and the unknown truth. For example, choosing a Gaussian probability distribution of zero mean for the perturbation will lead to a set of model realizations which symmetrically deviate from the interpolated reference model. The spatial frequency of the perturbation is typically described by a variogram (or spatial covariance) model. Both the distribution and the variogram model have a strong impact on the produced realizations and should be calibrated from available data and prior geological knowledge. It is also possible to account for uncertainty in the variogram range or in the dispersion of the distribution. In all cases, the estimation variance produced by kriging provides local estimation errors, which depend on the chosen variogram model and on the data layout. This information may be used directly in its discretized form (Fig. 10C) for some applications, but most works prefer generating realizations which produce possible models.

An early but very complete example of these approaches is available in Abrahamsen et al. (1991) and Abrahamsen (1993): they combined uncertainties in seismic time-to-depth conversion by kriging multiple subparallel geological horizons in a "layer-cake" model using a Bayesian kriging approach (Omre, 1987). The work of these authors denotes a step-change in the consideration of uncertainties, as they also explicitly consider the correlation of multiple layers through the seismic input data. They extended the work subsequently further to hydrocarbon-in-place estimates (Abrahamsen, Lia, & Omre, 1992), gross volume estimates (Abrahamsen, Hauge, Heggland, & Mostad, 2000), and the management of inequality constraints (Abrahamsen, Dahle, Hauge, Almendral-Vazquez, & Vigsnes, 2015).

Another example of this approach is the probability-field method (e.g., Lecour, Cognot, Duvinage, Thore, & Dulac, 2001; Thore et al., 2002), which can be applied to geological interfaces and faults as surface elements, even in complex 3-D settings. A first idea in this case is to consider perturbation amplitude not along the vertical direction but orthogonally to each surface. The knowledge of the contacts between surfaces is used to define in which order the various structural surfaces are perturbed. For example, in the simple case of a horizon cut by a fault, the idea is to first perturb the fault geometry, then the horizon geometry to maintain the maintain the contact between both surfaces (Lecour et al., 2001). The second idea is to precompute the range of uncertainty around each surface by carefully propagating seismic imaging uncertainties and interpretation uncertainties. The uncertainty envelopes defined around each surface implicitly define a cumulative distribution function (cdf) along the surface normal (very similar to the representation in Fig. 11B). This cdf is then sampled by generating a spatially correlated random field between 0 and 1 using the probability-field method (Srivastava, 1992). As the envelope already reflects data uncertainties, this method provides an alternative to conventional conditional simulation procedures with the advantage that fast spectral methods can be implemented to simulate the probability fields (e.g., Pyrcz & White, 2015).

These methods have been widely applied (e.g., Charles, Guemene, Corre, Vincent, & Dubrule, 2001; Corre, Thore, de Feraudy, & Vincent, 2000; Irving, Chavanne, Faure, Buffet, & Barber, 2010; Samson et al., 1996; Suzuki et al., 2008) to generate models sampling uncertainties both around horizons and faults. (Holden et al., 2003; Hollund et al., 2002; Lecour et al., 2001; Rivenæs, Otterlei, Zachariassen, Dart, & Sjøholm, 2005; Thore et al., 2002). Some authors mention, thought, that the method can create geologically unrealistic surfaces (e.g., Pyrcz & Deutsch, 2014a; Srivastava & Froidevaux, 2005). In the case of faults, recent improvements have been proposed by Røe, Abrahamsen, Georgsen, Syversveen, and Lia (2010) and Røe, Georgsen, and Abrahamsen (2014), treating faults as surfaces in a rotated coordinate system and calculating horizon displacements using 3-D vector fields (see Georgsen et al., 2012; Hoffman & Neave, 2007b).

These methods are, in general, limited to relatively simple structural settings not affected by faults (Fig. 2A), or, in faulted domains (Fig. 2B), to cases where geometric uncertainties are small relative to fault spacing. Indeed, the sequential perturbation of many interfaces is very difficult to implement.

4.1.3 Geometric model perturbation: Space warping methods

In geological contexts where many surfaces are present, the surface perturbation strategy relies on the careful consideration of the perturbation order to maintain model consistency. In the case of small perturbations, a volumetric approach can be applied instead. The principle is the same as for surface perturbation, but it uses a three-dimensional vector field instead of a scalar field applied in a prescribed direction. The main advantage is to consider all geomodel elements at once, treated as embedded in an elastic material. This is proposed for instance by Caumon, Tertois, and Zhang (2007), who combine several two-dimensional perturbation fields generated independently on a set of faults to generate a global three-dimensional scalar field. As the input vector fields may cross, a divergence-free term is added to the interpolation to ensure that the perturbation is volumetrically consistent. This methodology was used for instance by Suzuki et al. (2008) to create a population of structural models by direct deformation of a hexahedral reservoir grid.

In the context of local model updating, Tertois and Mallet (2007) propose a method to bound the variation of volume throughout deformation in the context of interactive model editing by experts. This approach was used by Caumon et al. (2007) and Mallet and Tertois (2010) to globally deform a model to reflect some small geometric fault uncertainties. More recently, Laurent, Caumon, and Jessell (2015) proposed to adapt an "as rigid as possible" method from computer graphics to also interpolate a 3-D displacement field between a set of fixed points. This type of method could certainly be extended to deform a reference model volumetrically. However, as large model distortions may occur in the case of large perturbations, these methods are only applicable for relatively mild uncertainties.

4.1.4 Geometric model perturbation: Implicit methods (implicit m, κ, and β)

In domains of high complexity (Fig. 2B and C), extending implicit interpolation methods (Section 3.2) has significantly helped accounting for uncertainties. This includes methods which perturb existing data or simulate synthetic data points (Jessell, Ailleres, & Kemp, 2010; Lindsay et al., 2012; Lindsay, Perrouty, Jessell, & Ailleres, 2013; Wellmann et al., 2010a) and also methods which perturb the implicit scalar field away from the data (Caumon, 2010; Caumon et al., 2007; Cherpeau & Caumon, 2015; Mallet & Tertois, 2010).

In the potential-field method (see Section 3.2.3 and Calcagno et al., 2008; Lajaunie et al., 1997), the use of kriging algorithm provides, like in map-based approaches, a direct estimate of interpolation uncertainty through estimation variance. This approach has been implemented by Chilès et al. (2004) and Aug et al. (2005), but the calibration of the variogram model relies on the availability of many orientation measurements. However, the possibility to obtain a direct measure of uncertainty in the interpolation step is a promising aspect for future research.

The direct link to the initial geological input data also allows a direct consideration of effects of data density. This possibility has been used by Putz et al. (2006) to evaluate the effect of reduced data density on a geological model, generated mostly from field observations in the Eastern Alps. The potential of this approach to consider uncertainties related to data positions in poly-deformed terranes has also been recognized by Maxelon et al. (2009), especially as many interface points have to be placed to constrain subsurface structures, where only very limited or no direct observation is available.

An important reason for the direct consideration of uncertainties in interface points is the possibility to treat high uncertainties related to the common requirement to include virtual data points to regulate model geometry, especially at depth when no abundant information is available. This is especially the case when creating models in regions where no high-quality seismic or borehole data is available, for example, in regional-scale geological modeling (Jessell et al., 2014). These additional points are then commonly placed on the basis of expert knowledge, and therefore highly subjective and potentially also highly uncertain.

In mesh-based implicit methods, the kriging formalism is generally replaced by a discrete formulation of thin-plate spline energy or some other roughness criterion (see Section 3.2.3 and Caumon, Gray, et al., 2013; Frank et al., 2007; Laurent, 2016; Moyen et al., 2004; Souche et al., 2013). Independently of how exactly a mesh-based implicit model is computed, a simple and direct way to perturb the model away from the observations is to add a spatially correlated random field $\epsilon(\mathbf{x})$ to the reference implicit scalar field $s(\mathbf{x})$ (Aydin & Caers, 2017; Caumon, 2010; Caumon et al., 2007; Cherpeau & Caumon, 2015; Cherpeau et al., 2010; Mallet & Tertois, 2010). In this approach (Fig. 12), the random field values should be constrained to be zero at (certain) data locations. Away from the data, the random field must vary smoothly to avoid introducing kinks in the perturbed surface geometry. Scaling the random field $\epsilon(\mathbf{x})$ by $1/||\nabla s(\mathbf{x})||$ approximates the horizon shift induced by the perturbation. In the presence

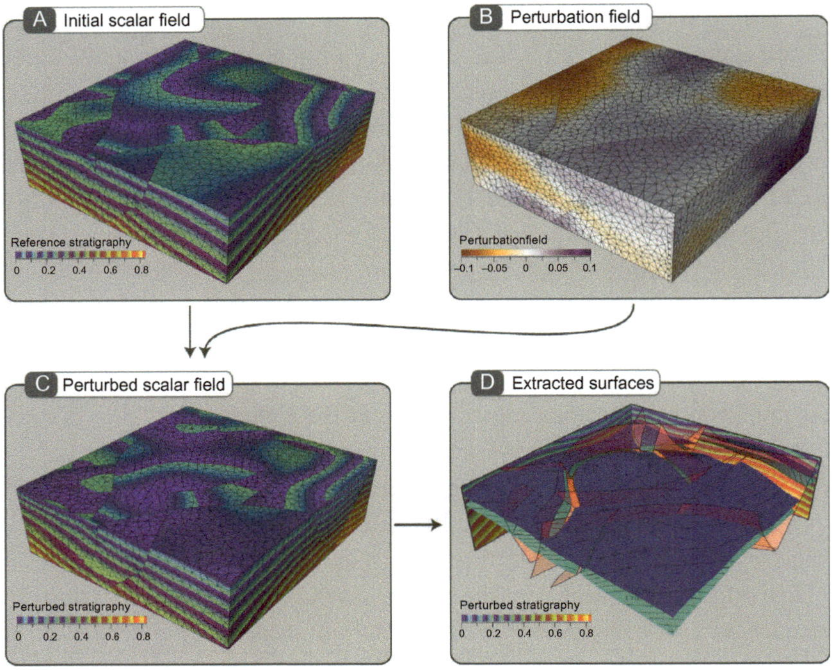

Fig. 12 Perturbing an implicit stratigraphic model: (A) Initial scalar field $s(\mathbf{x})$ representing relative geological time. (B) Random field $\epsilon(\mathbf{x})$ used to perturb the stratigraphy. (C) Perturbed scalar field $s'(\mathbf{x}) = s(\mathbf{x}) + \epsilon(\mathbf{x})$. (D) View of two perturbed horizons; faults are displayed in *semitransparent red* and the perturbed scalar field is displayed on the *background planes.*

of faults, Caumon et al. (2007) suggest that computing the perturbation field in depositional space (Mallet, 2004) preserves the fault slip. Cherpeau and Caumon (2015) used this idea to produce a relatively large continuous perturbation across faults (preserving the fault slip) with a relatively smaller discontinuous perturbation (changing the fault slip). Further approaches to geometrical perturbations of implicit stratigraphic models are also discussed by Mallet (2014, Chapter 9).

In all cases, implicit methods do not explicitly allow for controlling the connectivity of the level-set surfaces. A consequence is that adding virtual data points or adding a random spatially correlated noise $\epsilon(\mathbf{x})$ may change the topology of the surfaces by creating closed iso-surfaces. This possibility may be a strength for some complex geological bodies such as ore bodies, as their connectivity can be uncertain; conversely, it may generate inconsistent models in the case of stratigraphic formations, as depositional processes can never generate closed surfaces.

Interesting extensions are also the recent developments of implicit methods for modeling of multistage fold geometries and overprinting deformation by Laurent et al. (2016), which allows an integration of these structural elements into uncertainty quantification and inverse frameworks (Grose et al., 2017, 2018). In essence, this approach infers the parameters of Fourier series describing fold limb rotation and fold axis rotation angles from structural observations. The obtained posterior distribution are then used to estimate the uncertainty in poly-phased implicit fault models.

4.1.5 Topological uncertainties and data association (m, Φ, κ, β, and τ)

The above model perturbation methods allow to sample uncertainty around a particular structural interpretation. Depending on the geological context, these perturbations may imply a few topological changes due to truncation by faults, unconformities, or intrusions (see Section 3.1.1 and Fig 7). Indeed, the geometric perturbation followed by the truncation of a particular geological interface may change the number of connected components or the number of holes in this interface. In the example of Fig. 12, this could happen for instance when an implicit horizon goes above or below the branch line formed by X-shaped, λ-shaped, or Y-shaped contacts between fault surfaces.

However, these changes remain relatively limited, in the sense that they do not significantly change the interpretation. More drastic topological changes imply considering variable numbers of geological interfaces, various ways to associate observations, and possibly also various relative ages (and truncation rules) for these interfaces.

In the context of fault uncertainty management in petroleum reservoirs, Munthe, Holden, Mostad, and Townsend (1994), Hollund et al. (2002), and Holden et al. (2003) pioneered by formalizing the main components of uncertain fault models. Their stochastic fault model is based on fault objects which are simulated using a marked point process. Prior information is translated using notions of fault families characterized by size, orientation, and displacement distributions. Each fault is simulated by a stochastic point process which may include attraction or repulsion with previously simulated faults or fault families. Fault may also be truncated in the simulation process based on geometrical considerations. Last, each fault generates a discontinuous displacement field shifting the layers on either side of the fault. This approach has been mainly applied on extrusion-based 2.5D representations of subsurface models, see, for example, Rivenæs et al. (2005).

As the representativity of these models is limited, more recent work has focused on extending these methods to more general three-dimensional cases (Aydin & Caers, 2017; Cherpeau & Caumon, 2015; Cherpeau et al., 2010; Georgsen et al., 2012; Maerten, Gillespie, & Daniel, 2006; Røe et al., 2010). Another recent area of research concerns the conditioning to observation data. Indeed, stochastic fault models may be seen as a way to associate spatial data (e.g., points where faults have been observed) consistently with prior knowledge (e.g., ideas about the fault family orientations, sizes). However, generating fault using a spatial point process until all observations are matched can be very inefficient. Therefore, Cherpeau et al. (2010) and Cherpeau and Caumon (2015) have proposed a method where fault evidence are associated sequentially by sampling fault centers near the data and then trying to associate other fault data consistently with prior information. This problem has recently been formalized using a graph which can be sampled to generate stochastic fault networks (Godefroy, 2018).

Similar methods have also been applied in stratigraphic settings to simulate the possible location of stratigraphic unconformities from sparse borehole data (Edwards et al., 2018; Lallier et al., 2012, 2016; Wu & Caumon, 2017). As the location and number of gaps in the stratigraphic record vary, this amounts to changing the number of units and relationships between units in the stratigraphic column.

4.1.6 Alternative modeling approaches

In addition to the approaches mentioned above, several variants for stochastic model generation exist for alternative modeling approaches. Several authors used simple parametric models of lines and planes to evaluate uncertainties of faults, for example, the propagation at depth related to uncertain observations and measurements at the surface using linear projection functions (e.g., Bistacchi et al., 2008) or to link fault and shear zone observations at the surface and in an underground lab (Schneeberger et al., 2017). In our notation, these approaches can be modeled by simple linear basis functions for linear regression, where position and gradient can be adjusted through the coefficient vector $\boldsymbol{m}$.

Modeling approaches based on kinematic modeling concepts are difficult to describe with the notation in Eq. (15), as they are based on modeling the interaction of multiple geological events with kinematic equations (e.g., Jessell, 1981; Jessell et al., 2014). However, this approach has also been used in a stochastic framework to estimate the effect of uncertain kinematic parameters (e.g., fault geometry, folding parameters) and timing of geological events (e.g., the order of faults) using the automatic modeling implementation pynoddy (Jessell et al., 2014; Wellmann et al., 2016).

4.2 Dealing with multiple models: Visualization and communication of uncertainties

There are generally two reasons for evaluating uncertainties in a study (e.g., Beven, 2016): (1) to answer a scientific question, improve understanding and evaluate different hypotheses, and (2) as a guidance for a subsequent decision-making processes. In the second case, the communication of uncertainties becomes an elementary aspect. As these visualizations form an important basis for decisions (often by nonexperts and stakeholders with a varying background and expertise in the topic), the uncertainties associated with these models should be communicated.

The analysis and representation of uncertainties in a spatial context is in its essence different to common approaches in uncertainty quantification approaches, where often the statistical distribution of one or several parameters is already sufficient as a measure of uncertainty (e.g., Gelman et al., 1995; MacKay, 2003). In the spatial context, the main interest is often to determine the uncertainty of a model outcome (geological unit, porosity, etc.) at *all* locations in the investigated domain. The following description is therefore specifically relevant to these cases where the representation of the geological model is actually already a final aim, and it is not directly processed further to, for example, a process simulation, or an analysis which would lead to a dimension reduction (e.g., calculations of volume of a resource, geothermal heat in place, the outcome at a previous specified single location). This is to say, in all cases where the subsequent model use is not directly obvious or where the spatial structure itself will form part of a decision progress (e.g., in well planning, planning for next exploration steps or experimental design for additional measurements).

The visualization and communication of uncertainties is an active field of research (see, for example, the excellent overview by Spiegelhalter, Pearson, & Short, 2011) and, in the spatial context, it has long been driven by approaches to visualizations in a geographic context of maps (Goodchild, Buttenfield, & Wood, 1994; Goodchild, Chih-Chang, & Leung, 1994; Leung, Goodchild, & Lin, 1993; MacEachren, 1992; MacEachren et al., 2005; Pang, Wittenbrink, & Lodha, 1997). Applications of these concepts have since been successfully applied in various geoscientific applications, notably in the analysis of geohazards (e.g., Kunz, Grêt-Regamey, & Hurni, 2011; Pang, 2008), with specific applications to flood prediction (e.g., Beven, Lamb, Leedal, & Hunter, 2014; Bruen et al., 2010; Lim, Brandt, & Seipel, 2015; Seipel & Lim, 2017), tsunami risk mapping (e.g., Goda & Song, 2015), hurricane prediction (e.g., Cox, House, & Lindell, 2013), seismic risks (e.g., Bostrom, Anselin, & Farris, 2008), as well as in different analyses of remote sensing scenes (Comber, Fisher, Brunsdon, & Khmag, 2012; Van der Wel, Van der Gaag, & Gorte, 1998) and in hydrogeological models (e.g., Benke, Pettit, & Lowell, 2011). But even though many approaches to visualize and communicate uncertainties have been developed in these related fields, most have been developed for the 2-D context of maps and sections and they are partly not directly applicable to the full 3-D setting of geological models considered here.

We can generally distinguish two approaches to visualize and communicate uncertainties from ensembles of generated models:

1. the visualization of multiple model realizations, and methods for *morphing* between different model realizations in a movie-form, and
2. the quantitative analysis and subsequent visualization of uncertainties using summary measures (e.g., variance, entropy, etc.).

These two aspects are described in the following.

4.2.1 Simultaneous representation of multiple models

An intuitive and direct visualization of spatial uncertainties is the representation of multiple possible model outcomes. This method has long been a standard method to represent uncertainties in geostatistical studies (e.g., Mariethoz & Caers, 2015; Pyrcz & White, 2015) and to visualize multiple outcomes from geophysical inversions (e.g., Boschetti & Moresi, 2001; Mosegaard & Tarantola, 1995). Also in the field of geological modeling, this type of representation is widely used (e.g., Jessell et al., 2010; Lindsay et al., 2012; Wellmann et al., 2010a). A typical example is the presentation of multiple reservoir models, for example, in Suzuki et al. (2008), presented in Fig. 13A. These representations provide a good intuition for a possible variability in the generated models, but it quickly becomes difficult to interpret where exactly changes in the model occur and the representation is, for practical reasons, restricted to a limited number of figures.

A representation of multiple model outcomes in one figure provides a more direct representation of this variability. An example for multiple models in one section is given in Fig. 13B from De la Varga et al. (2018), where the spatial variability for three different layers is clearly visible. This type of representation has also been attempted in 3-D, an example by Mallet and Tertois (2010) is presented in Fig. 13C. However, it is obvious that this

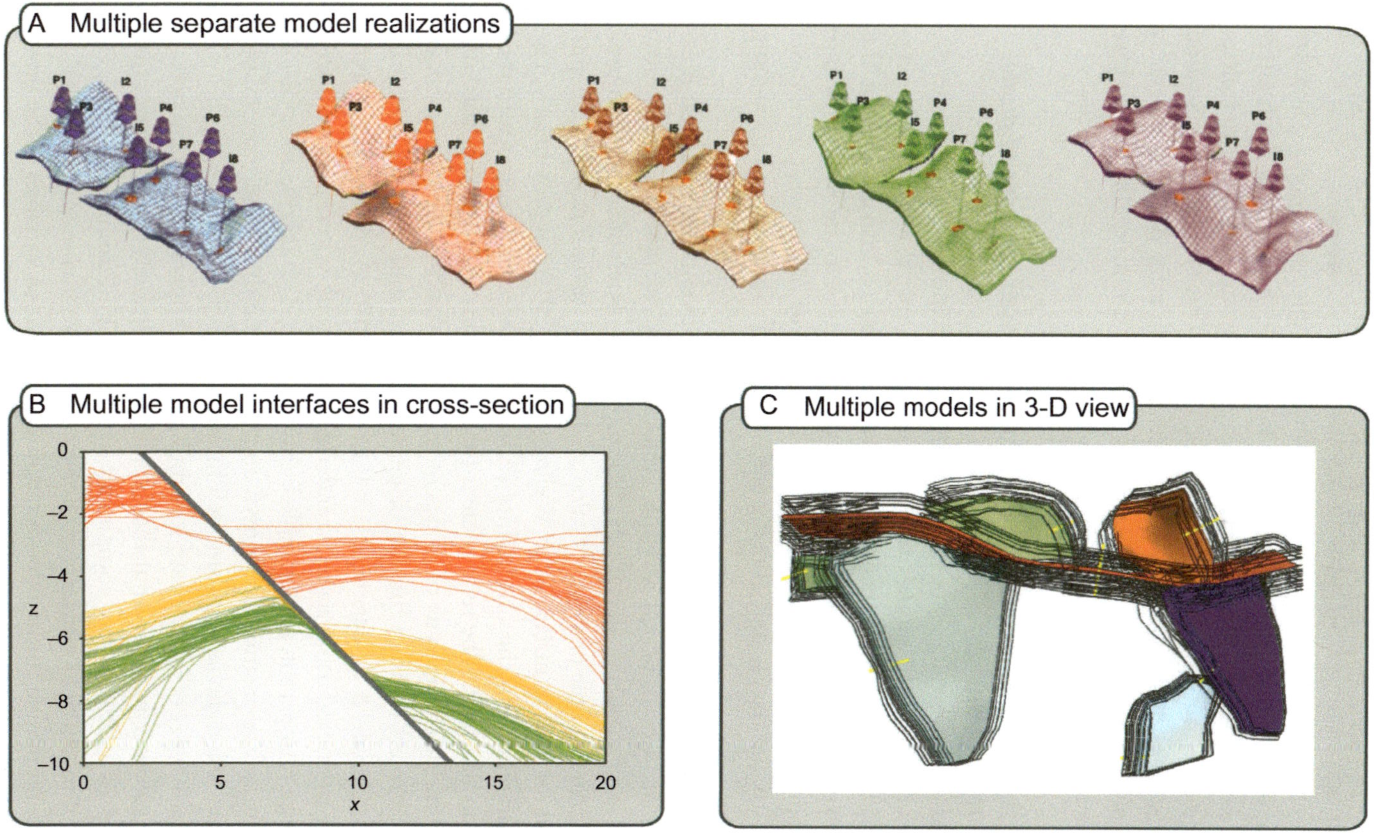

Fig. 13 Visual representation of multiple model realizations: (A) Model realzations in separate adjacent subfigures (Suzuki et al., 2008). (B) Simultaneous representation of multiple interfaces in a 2-D section (De la Varga et al., 2018), and (C) in a 3-D representation (Mallet & Tertois, 2010).

type of representation can quickly get confusing and difficult to interpret for increasingly complex 3-D structural settings.

A related possibility is the visualization of multiple models on a computer screen, either as a defined succession in the form of a movie, or in an interactive representation, where the user can iterate through a number of realizations (e.g., Kunz et al., 2011; Srivastava, 1994). This type of representation resolves some of the problems of a presentation in a fixed image, especially if it is possible to also rotate views in full 3-D, but it is more difficult to share or to integrate into a publication or report.

4.2.2 Approaches to quantify uncertainties based on a set of model realizations

Instead of the simultaneous representation, an additional approach is to use the set of generated models first to determine a spatial estimate of uncertainty, and then to determine a suitable method to visually present this uncertainty. Approaches to quantify uncertainty can be separated into methods directly using continuous measures (layer thickness at a location, depth to interface), and those using categorical values (geological unit, lithology class, etc.).

For continuous variables, a wide range of conventional statistical measures can directly be applied and visualized, for example, showing mean values and standard deviations (Potter, Kniss, Riesenfeld, & Johnson, 2010). An example of a map of calculated standard deviations of layer depth, draped on a mean surface from Wellmann et al. (2010a) is presented in Fig. 14A, highlighting the uncertainty in layer depth related to the uncertain position of a fault. The requirement for realizations to follow a normal distribution for the analysis of values of mean and standard deviations can be circumvented through the use of percentiles and min/max surfaces (e.g., Potter, Gerber, & Anderson, 2013; Røe et al., 2014). A similar visualization can be attempted using volume rendering techniques, where uncertain areas are more transparent. An example is shown in Fig. 14B. Here, uncertainties of an interface position are shown and it is visible that these uncertainties increase when deviating from well locations.

These types of analyses are straightforward to apply and implemented in many software packages, but their application is limited to 2.5-D structural settings of relatively low complexity (see Fig. 1) and cannot easily be generalized to full 3-D settings (see, for example, Wellmann et al., 2010a).

An alternative to the consideration of continuous variables is the direct use of the categorical value of a geological unit or lithology in 3-D space.

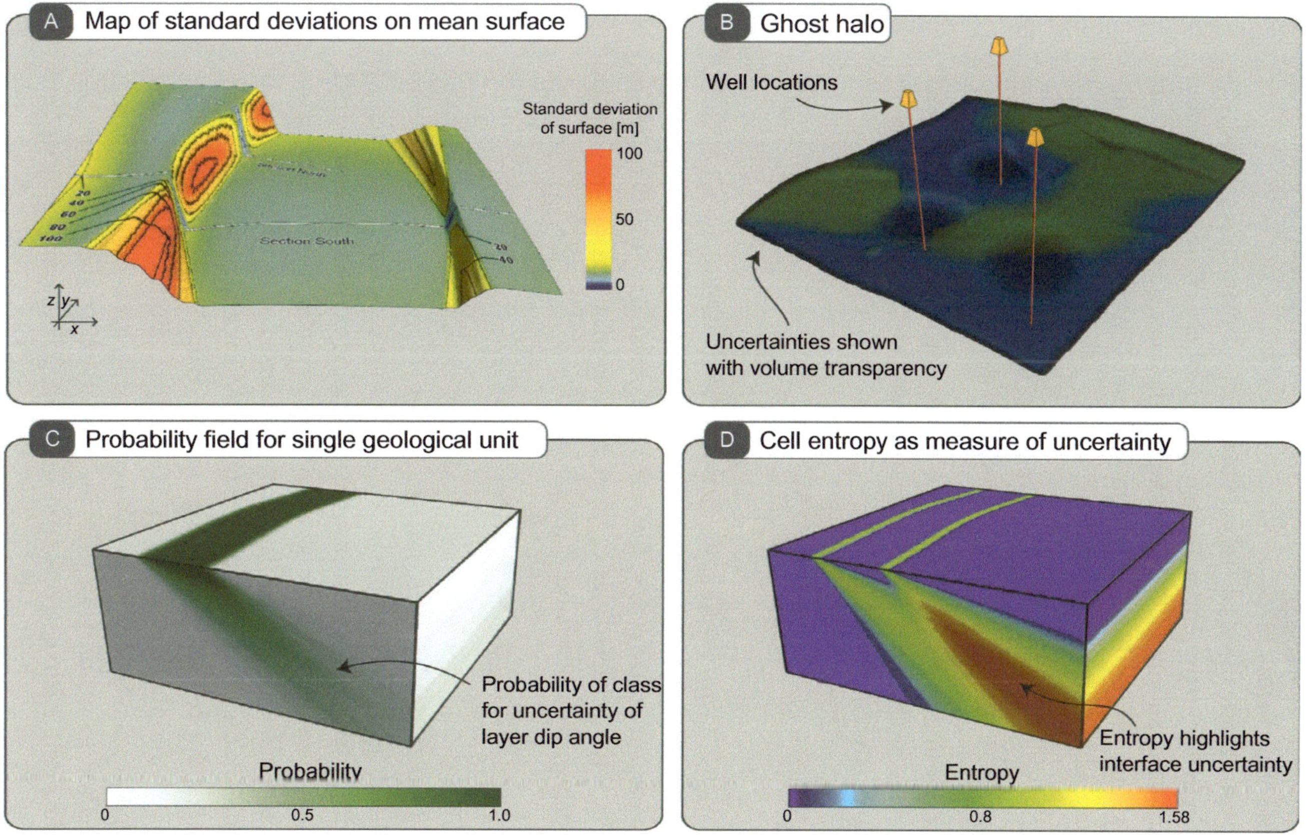

Fig. 14 Uncertainty visualization on the basis of a scalar quantity of uncertainty: (A) Standard deviation of thickness in map view draped on mean surface (Wellmann et al., 2010a). (B) Ghost halo as a volumetric vizualisation of surface uncertainty (T. Viard, personal communication) (C) Calculated cell-based probability for a specific geological unit, and (D) cell entropy as combined measure of uncertainty (Wellmann & Regenauer-Lieb, 2012).

The following description of this type of spatial quantification of uncertainties follows Wellmann et al. (2010a), Wellmann and Regenauer-Lieb (2012), and Wellmann (2013).

We consider here the subdivision of the model domain into defined subspaces, where $m(\mathbf{x})$ represents a discrete geological class C (e.g., a lithology type or stratigraphic unit).[i] We can identify the class membership at each location $\mathbf{x}$ and, on this basis, define an indicator function for each model i in the ensemble of n generated models:

$$\mathbb{1}_C(\mathbf{x}, i) = \begin{cases} 1 & \text{if } \mathbf{x} \in C \text{ in model } i \\ 0 & \text{if } \mathbf{x} \notin C \text{ in model } i \end{cases} \tag{16}$$

This indicator function is then used to estimate a probability $P_C(\mathbf{x})$ for each location in the model domain to belong to a class C on the basis of an ensemble of n generated models:

$$P_C(\mathbf{x}) = \frac{1}{n} \sum_i \mathbb{1}_C(\mathbf{x}, i) \tag{17}$$

In this way, we obtain a multinomial probability field (see also Goodchild, Buttenfield, & Wood, 1994), which can be used to visualize the probability for each separate possible class C (i.e., each geological unit) in 3-D representations (Fig. 14C). The variance $\sigma_C(\mathbf{x})$ of the categorical variable may also be used to map spatial uncertainty:

$$\sigma_C^2(\mathbf{x}) = P_C(\mathbf{x})(1 - P_C(\mathbf{x})).$$

This analysis and representation are directly suitable when we are interested in a single possible outcome (e.g., a reservoir layer or a mineable resource), but it is less useful as an overall measure of model uncertainty if more than two geological formations or lithology classes are considered, as we would then need multiple figures to present probabilities for each class.

For these cases, a suitable summary measure for a representation of uncertainties in the full 3-D space is required. Motivated by previous applications using the definition of spatial entropy (Batty, 1974) and the representation of uncertainties in maps (Goodchild, Buttenfield, & Wood, 1994; Leung et al., 1993), the use of cell entropy has been shown to be a useful method to quantify and visualize uncertainties about the full 3-D structural model (Wellmann & Regenauer-Lieb, 2012). It has similarly been

[i] Note that this subdivision is not necessarily identical to the separation into regions R_J as in Section 3.1.1, as multiple regions can belong to the same class C.

applied to wireline log classifications (Grana et al., 2012) and cross sections (e.g., Elfeki & Dekking, 2005). Specifically, we use the Shannon entropy (Shannon, 1948) to calculate an entropy value at each location **x** on the basis of the class probabilities P_C:

$$H(\mathbf{x}) = -\sum_{C} P_C(\mathbf{x}) \log_2 P_C(\mathbf{x}) \tag{18}$$

The advantage of this method is that we now obtain a scalar measure of uncertainty at each location **x** in space, and this aspect allows for a direct representation of model uncertainties in space.

Even though entropy values can be calculated in a continuous field, it is common practice to use a defined space-filling grid structure and to calculate entropy values for each cell for the purpose of visualization. A simple example for the use of cell entropy for 3-D geological models is presented in Fig. 14D (after Wellmann & Regenauer-Lieb, 2012). The difference to the class probability representation (Fig. 14C) is here clearly visible: whereas in the latter case, uncertainties about a specific class outcome are shown, entropy provides a cumulative measure of uncertainty. Entropy as a measure of uncertainty has been applied successfully in geological modeling studies, both for the representation of model uncertainties (e.g., Bianchi, Kearsey, & Kingdon, 2015; Schweizer et al., 2017), as well as the comparison of uncertainty reduction in the context of Bayesian inversion (e.g., De la Varga et al., 2018; Wang, Wellmann, Li, Wang, & Liang, 2017; Wellmann, de la Varga, Murdie, Gessner, & Jessell, 2017).

Apart from the use as a measure to visualize uncertainties, it should also be noted that the value itself has a clear quantitative meaning: the upper bound is defined by the number of classes k

$$H_{\max} = \log_2 k \tag{19}$$

and this value is obtained when all class outcomes are equally likely, corresponding to the case of maximum uncertainty. For more details on the underlying theoretical concepts, see Cover and Thomas (2005).

Additional aspects about the quantitative spatial interpretation, the calculation of related fuzziness measures to quantify uncertainties of specific classes, as well as model average values is described in Wellmann and Regenauer-Lieb (2012). Furthermore, measures from information theory can be applied to determine the spatial correlation of uncertainties (see Wellmann, 2013).

Another concept to quantify uncertainties in geological models on the basis of class membership is the measure of stratigraphic possibility, as the number of possible (lithology) classes $L(\mathbf{x})$, and variability $V(\mathbf{x})$ at each location, introduced by Lindsay et al. (2012). In the notation used above, this measure can be calculated as:

$$V(\mathbf{x}) = 1 - P_{\hat{C}}(\mathbf{x}) \quad (20)$$

where $\hat{C}$ is the class with the highest estimated probability at position $\mathbf{x}$. Both measures should be used for a representation of model uncertainty. For a more detailed description and examples, see Lindsay et al. (2012) and Lindsay, Perrouty, Jessell, and Ailleres (2014).

4.2.3 Communication of uncertainties

With the methods described above, we obtain a scalar measure of uncertainty for the model domain, which can be visually represented in 2-D and 3-D views. However, the general communication of these uncertainties is, in itself, a challenging aspect. The question of uncertainty visualization is completely entwined with the question of how uncertainties are perceived by potential stakeholders—and therefore on psychological aspects (e.g., Spiegelhalter et al., 2011). Certainly, the choice for the best method to represent uncertainties depends on the specific type of data, the context, and the target audience (see Visschers, Meertens, Passchier, & De Vries, 2009). One possibility would be the *adjacent representation* of an uncertainty model next to a single model representation (Viard, Caumon, & Levy, 2011). But many studies showed that even the communication of probabilities to lay audiences can be difficult (e.g., Spiegelhalter et al., 2011), and it will not be easier with the derived concepts of entropy or stratigraphic variability.

Significant research has gone into the *coincident representation* of uncertainties, i.e., the integrated representation of uncertainty together with primary data or a model, for a more direct communication of uncertainties. MacEachren (1992) and Pang et al. (1997) describe a variety of methods to represent uncertainties of (continuous and categorical) scalar, vector, and tensor data. These approaches can be classified into intrinsic approaches (changing object appearance) and extrinsic methods (adding symbols or features to provide uncertainty information). Intrinsic methods include the use of hue and saturation of color, blurring, and focus to distinguish uncertain model areas (e.g., Djurcilov, Kim, Lermusiaux, & Pang, 2002, 2001; MacEachren, 1992; MacEachren et al., 2005; Seipel & Lim, 2017). There have also been studies on the interactive visualization (e.g., Kunz et al., 2011) and animated visualizations (e.g., Bostrom et al., 2008), even including vibrations in regions of uncertainty (e.g., Brown, 2004).

Finally, we would like to mention that there are also approaches to represent uncertainties in geological models, which are not based on the analysis of a model ensemble, as considered here. These methods include plots of spatial data density (e.g., maps of kriging variance or similar approaches representing the distance to data), possibly combined with data accuracy (e.g., Berg, Mathers, Kessler, & Keefer, 2011).

Overall, it is important to keep in mind that a representation of uncertainty should be adjusted to the expected target audience. This includes the question if accuracy or uncertainty should be represented (e.g. Buttenfield, 1993), an intrinsic or extrinsic representation, and the specific type of uncertainty visualization. Overall, it can be said that, for a coincident representation of uncertainty, transparency has been found to be an intuitive measure, even though it is difficult to visualize (Viard et al., 2011), but more research into the perception of uncertainties in geological models is still required.

4.3 Uncertainty reduction

We described above methods to generate randomized realizations of geological models. Depending on the suitability of the prior distribution and the model structure itself, a subset of these realizations could be violating basic geological principles or simply lead to a structure, which is not conforming to the expected geological setting. In this case, we aim to define geological criteria to evaluate the validity of a generated model realization (Fig. 10D). In a similar sense, geological models can be related to geophysical measurements through the additional use of rock physics models (Fig. 10E). Examples are density values assigned to specific lithological classes, which allow a comparison to measured gravity anomalies, acoustic impedance for a comparison to seismic measurements, and permeability for a comparison to flow and transport measurements. In all of these cases, criteria can be defined to describe the validity of the generated model, and even to consider the geological model as part of a geophysical inverse approach.

A related task to the evaluation of model results with these additional geological and geophysical criteria is to sort the generated set of models into specific classes, or to pick out "extreme" cases, to evaluate their effect in subsequent process simulations or predictions. It is often straightforward for a structural geologist to perform such tasks for a reasonable set of models. However, in order to scale this process up to a large set of realizations and to perform this process automatically, we require clearly defined methods for a quantitative comparison of models to other models, and of model results to additionally available geological and geophysical information, either in the form of measurements, observations, or even general aspects of geological knowledge.

We can here distinguish two different, but related, types of measures:

- Distance functions: measures of distance *between* model realizations;
- Mistfit functions: measures of distance from a model realization to an *external* additional observation or measurement.

Both aspects are related, as some measures could theoretically be applied in both contexts (e.g., a specific outcome at a given location could be used for model comparison, but also to define a misfit to an actual observation at this location). In the following, we first describe methods based on geologic criteria (Fig. 10D), before providing an overview of the link to geophysical criteria (Fig. 10E).

4.3.1 Model analysis and model comparison based on geologic criteria

Measures for model comparison can be directly based on the result of the interpolation function $m(\boldsymbol{x})$. In addition, values at multiple locations are often combined in an additional function, for example, to estimate summary measures or geometric aspects. Jessell et al. (2010) already provided a detailed list of misfit functions that can be applied to determine a distance metric between a model realization and diverse types of geological information, including spatial aspects of geological observations (point and volume information), as well as lineations and topological measures.

Especially the aspect of topology, as described by Jessell et al. (2010), is interesting as a measure of misfit, as topological relationships are often of practical interest (Pouliot, Bédard, Kirkwood, & Lachance, 2008). Further descriptions of topology in geology, as well as measures for the comparison of topological graphs are described in Thiele, Jessell, Lindsay, Ogarko, et al. (2016) and Thiele, Jessell, Lindsay, Wellmann, and Pakyuz-Charrier (2016). This research has also shown that topological graphs of geological structures quickly become very complex in full 3-D settings. But the abstraction to this level provides interesting possibilities to use this information as geological criteria to validate geological model realizations.

Additional geodiversity metrics are described by Lindsay, Jessell, et al. (2013) with the aim of model comparison. The authors extend the analysis to formation depth and curvature, and additionally include measures that are calculated on discretized representations of a geological model, such as volume and neighborhood relationships, to evaluate the juxtaposition of stratigraphic units, as an element of topology. An interesting aspect is also that the authors combine all metrics and use a principal component analysis to determine outliers and barycenter examples in the set of generated models.

Pellerin et al. (2015) describe measures of complexity for geomodel representations in boundary models, based on definitions from the field of geometric modeling and visualization (Rossignac, 2005; Sukumar, Page, Koschan, & Abidi, 2008). The authors distinguish structural model components, such as the number of surfaces and corners in a grid and local statistical neighborhood measures and evaluate the use of these measures to define a complexity space to classify models by their complexity. Several measures overlap with the definitions in Jessell et al. (2010) and Lindsay, Jessell, et al. (2013), but are here focused on the use for models in triangulated boundary representations.

Model dissimilarity has also been investigated by Schweizer et al. (2017) in an assessment of model scenarios with increasing complexity. The authors also evaluated the use of the Jaccard similarity measure and the extension to consider (cell) probabilities with the normalized city-block distance. In the application to a geological data set, the authors found that especially the city-block distance provides a suitable measure for model dissimilarity, especially in the combination with a spatial visualization of uncertainties using Shannon entropy, as described in Section 4.2.2. Related approaches have also been taken in the field of reservoir modeling and history matching (e.g., Park, Scheidt, Fenwick, Boucher, & Caers, 2013; Scheidt & Caers, 2008; Suzuki et al., 2008) with distance measures for production forecasts. In addition to these measures, an interesting path for future research is the investigation of more abstract methods to compress model space. For example, discrete cosine transforms (DCT) have successfully been used by Lochbühler, Vrugt, Sadegh, and Linde (2015) to compress information from multiple training images in sparse representations.

Similar to Jessell et al. (2010) and De la Varga and Wellmann (2016) evaluated the use of additional geological information to evaluate model realizations, but with a focus on the defintion of likelihood functions in a Bayesian inference framework (see Appendix A.1), where this additional geological information is considered as additional data and encoded in suitable likelihood functions. This approach has subsequently been combined with an optimized implementation of the implicit co-kriging interpolation (Calcagno et al., 2008; Lajaunie et al., 1997) in a probabilistic programming framework (De la Varga et al., 2018). As this method provides the possibility to generate model realizations directly on basis of the geological input parameters (Fig. 11A), this combination enables a probabilistic inference of geological model parameters, in addition to the optimization of suitable model realizations. This approach also opens-up the path for a full probabilistic joint inversion of geological and geophysical data (e.g., Wellmann et al., 2017).

A similar approach to combine different types of geological information was taken by Grose et al. (2018). The authors used a fold model to obtain posterior distributions for input parameters of an implicit fold modeling method (Laurent et al., 2016). In essence, the approach consists of a probabilistic optimization approach where parameters of a Fourier series describing fold limb rotation and fold axis rotation angles are inverted from structural observations. The obtained posterior distributions are then used to estimate the uncertainty in a subsequent modeling step, where spatial uncertainties are then visualized using Shannon entropy and an average angular distance to determine locations for uncertainty reduction. A similar approach, though not in a probabilistic framework, has been taken by Cardozo and Brandenburg (2014) in a kinematic inverse modeling of folding above propagating listric thrusts using simulated annealing to determine a range of models that fit observed structures.

We also note that these misfit functions, described in a general term in many publications, also provide the possibility to be included in a Bayesian framework in the form of likelihood functions, which can either be formally defined on the basis of measurement uncertainties or used in the form of summary measures using concepts of Approximate Bayesian Computation (Beaumont, Zhang, & Balding, 2002; Marjoram, Molitor, Plagnol, & Tavaré, 2003) or generalized likelihood functions (e.g., Beven & Binley, 2014). A brief introduction into the probabilistic viewpoint on uncertainty quantification is provided in Appendix A.1.

All these approaches show the relevance of determining suitable mesures to compare sets of model realizations and to find methods to integrate all available geological information in a modeling study. In this context, probabilistic approaches have recently shown promising results, and we expect further interesting work in this direction in the future.

4.3.2 Geological models and geophysical inversion

The combination of geophysical measurements and geological knowledge appears in many forms and contexts in geosciences. The fundamental problem of nonuniqueness in geophysical inverse problems is well known and has led to many developments to combine geophysical inversion with additional information since the early work of Backus and Gilbert (1967),

Parker (1974), and Tarantola and Valette (1982). Many recent developments address this issue through the use of multiple geophysical data sets in joint geophysical inverse approaches, and these approaches are described in detail in the geophysical literature (e.g., Gallardo & Meju, 2003, 2011; Moorkamp, Heincke, Jegen, Roberts, & Hobbs, 2011). We also mentioned important combinations of geological modeling and geophysical inversion in Section 4.1.2. Here, we consider more broadly approaches with important links between geological knowledge, encoded in 3-D geological models, and geophysical forward models and inversions.

Many authors have stated the requirement to combine geophysical and geological information. Fullagar et al. (2008) actually describe geophysical inversion as an extension of geological modeling, and Jessell et al. (2010) express the same view, from the viewpoint of geological modeling. Common approaches to include geological knowledge into geophysical inversions are the methods implemented in potential-field inversions. Nonuniqueness in the interpretation of gravity is already obvious from the fact that the effect of gravity can be formulated as a boundary value problem, and therefore all density distributions with the same boundary effect cannot be distinguished (e.g., Jacoby & Smilde, 2009). This leads to the problem that additional constraints and regularization parameters have to be defined in order to obtain a stable inversion result (e.g., Aster et al., 2005; Jacoby & Smilde, 2009; Telford et al., 2009). Additional depth-weighting methods can be applied to avoid a concentration of mass near the surface (e.g., Li & Oldenburg, 1998). The consideration of structural geological information from geological measurements is also explored in the work of Lelièvre and Oldenburg (2009). An interesting recent discussion on the problem of nonuniqueness in the interpretation of potential-field data has been contributed by Saltus and Blakely (2011). The authors directly state the relevance of prior knowledge for a successful interpretation of potential-field data and highlight that typical potential-field studies incorporate additional constraints to obtain reasonable solutions out of the infinite universe of possibilities.

The most widely used method to consider geological information in the form of geological models is the use of these models to initialize a subsurface parameterization for a geophysical inversion or process simulation (see also Jessell et al., 2014). All of these approaches are based on the concept that domains with similar properties can be defined in the subsurface, based on similar geological processes in the formation of a region. This concept is clearly related to the fundamental considerations for the use of structural models that we described at the beginning (Section 1.1), and directly integrated into the diverse interpolation formulations in Section 3.1 and the formulation of a physical value of interest $m(\boldsymbol{x})$ (e.g., Eq. (1)).

In order to obtain a meaningful model parameterization for geophysical inversion, relevant rock properties have to be assigned according to the lithological classes given by the geological model. We do not aim to describe the underlying concepts of rock physics for this step, but refer to the relevant literature (e.g., Mavko, Mukerji, & Dvorkin, 2009; Mukerji et al., 2001).

We will first consider geophysical inverse approaches based on the adjustment of geometric bodies. These methods are also referred to as the *geometric* property mode (following Silva et al., 2001). Many examples for this approach exist again for gravity and magnetic inversion and they are often based on 2.5-D assumptions of polygonal objects in a section with an infinite lateral extent, going back to the developments of Talwani, Worzel, and Landisman (1959) and Talwani (1965). This approach is, for example, implemented in the software GM-SYS (Won & Bevis, 1987) and IGMAS+ (Götze & Lahmeyer, 1988) and widely used for interactive geophysical modeling and inversion. These approaches often implement map- and extrusion-based explicit modeling approaches (Section 3.2.1), which also makes a combination with probabilistic geophysical inversions directly possible (e.g., Hauser et al., 2011; Hauser, Gunning, & Annetts, 2016). However, as described before, map- and extrusion-based approaches are limited to geological settings with a relatively low complexity (Fig. 2A).

Mesh-based geophysical inverse approaches in the *physical* property mode (Silva et al., 2001) are, instead, not limited to a specific complexity of the geological model, but require an initial discretization of the geological model into a grid structure (Fig. 10C). On this basis, geological models are often applied to obtain a subsurface parameterization for geophysical forward simulations and inversions. Still to date, many studies use geophysical forward simulations and then manually adjust the geological model based on a visual or quantitative comparison between forward model and geophysical measurements (e.g., Autin et al., 2016; Gradmann et al., 2013; Haase et al., 2017). However, discretized geological models have also long been used as a basis for geophysical inversion routines (e.g., Bosch, 1999; Bosch et al., 2001; Fullagar et al., 2008; Guillen et al., 2008; Moorkamp et al., 2011). Main differences in the approaches are mostly in the type of geological criteria that can be defined, for example, in the ability to fix specific known geometric objects, or to ensure neighborhood relationships between cells.

These concepts have been applied successfully in various gravity and magnetic studies (Calcagno et al., 2008; Joly, Chen, Faure, & Martelet,

2007; Joly, Martelet, Chen, & Faure, 2008; Lindsay, Perrouty, et al., 2013; Martelet et al., 2004; Wehr, Chevrot, Courrioux, & Guillen, 2018), but it has to be noted that the geological model is here merely an initial step to parameterize the geophysical inversion. It is not only possible, but potentially expected, that the final property distribution in the inverted model does not conform to the geological constraints and input data that were considered in the initial geological model used to obtain the parameterization (Jessell et al., 2014, 2010; Lindsay et al., 2014).

The problem of this loss of the link to the primary information in the geological model has recently been addressed by Giraud et al. (2017) combining uncertainty propagation with a probabilistic geological forward modeling method (Section 4.1) with joint geophysical inversions using a probability grid for each geological unit, instead of a fixed outcome in a cell (see also the calculation of probability grids in Section 4.2.2, Eq. (17)). This probability value is then combined with petrophysical models to obtain starting models and constraints for joint geophysical inversion and results suggest that the inversion routine properly considers areas of higher geological certainty. Related to this approach is the work by Wellmann et al. (2017), in which geological interface points of an implicit forward model are directly treated as parameters in a probabilistic inversion, and geological, as well as geophysical, likelihood functions are used to obtain posterior distributions of these geological input parameters. On the basis of the posterior distributions, multiple geometric model representations can then be obtained and further processed for uncertainty quantification and visualization.

As another recent example, Zheglova, Lelièvre, and Farquharson (2018) use implicit modeling of several geological interfaces coupled with petrophysical data for solving a joint gravity and seismic topography problem.

In the same spirit, Cherpeau, Caumon, Caers, and Levy (2012) propose to assess whether flow data can reduce uncertainties about fault locations and connectivity constrained by sparse geological cross sections and prior distributions of fault parameters for fault families. They use a sequential stochastic fault model which associates fault observations based on prior information. Using the random numbers used in this algorithm as parameter vectors, they run a Monte Carlo Markov chain (MCMC) algorithm to explore the parameter space. To include the possibility of having a different number of faults, they translate the prior probability of existence as modified distribution of fault size.

Other inverse models using a variable number of parameters are also proposed by Sambridge et al. (2013) and references therein. One of the key ideas is to use the number of model parameters as a variable in Bayesian inversion.

This method has been used for instance for layer-cake geological models with a variable number of layers represented by a one-dimensional Voronoi diagram. An interesting aspect of these transdimensional inversion methods is that they tend to favor the models which can explain observations with the smallest number of parameters (parsimonious solution).

An additional interesting path combining geophysical inversion with geological knowledge is the concept of *interactive inversion* (Boschetti & Moresi, 2001; Wijns, Boschetti, & Moresi, 2003; Wijns & Kowalczyk, 2007), which aims at an explicit integration of expert knowledge in the inversion process. This is achieved through a system generating possible scenarios. These scenarios are visualized on a screen, and experts select and rank feasible solutions. In a next step, a genetic algorithm is used to propose new realizations from these selected models. In a sense, this approach is between classical deterministic inversion with "trial-and-error" adjustments of settings (and similar approaches to evaluate the model space) and more integrated approaches, where model differences are included in the form of likelihood functions (e.g., Wellmann et al., 2017). An advantage of this interactive approach is that realizations can be produced and evaluated which have not been expected before (in difference to the defined likelihood functions). A potential disadvantage is the limited explicit statement by the expert on the decision, and therefore a limited reproducibility of results. However, the combination of geophysical inversion with a supervised machine learning approach is an interesting path to integrate different forms of geological knowledge, including tacit expert knowledge, in geophysical inversions.

5. Discussion and conclusions

Geological modeling methods have been rapidly developing over the last years—from improved representations in both explicit and implicit approaches, to more complex structural interactions and fault and fold models, to a more comprehensive consideration of uncertainties. With this review, we aimed to provide an overview of these developments and recent applications in the diverse fields using 3-D geological modeling approaches. One recurring aspect is the link to geophysical data processing and inversion approaches, both because geophysical data and interpretations are an essential input to conventional geological modeling studies, but also because geological concepts and models are increasingly used in joint geophysical and geological inversions.

Even though we provided an extensive description of 3-D structural geological models, the underlying concepts, methods, and uncertainties, we necessarily had to leave out several aspects in order to keep a focus on the most relevant topics, but also due to our own expertise. We have likely omitted to describe some related work, as it is barely possible to cover the entire breadth of such an interdisciplinary topic. Also, several of our descriptions are certainly open to discussion, as no single best-fitting and all-purpose geological modeling and uncertainty analysis method exists. In the following, we will contribute to this discussion and, finally, conclude with some suggestions and potential paths for future research in this field that we envisage for the coming years.

5.1 Discussion

5.1.1 On the two viewpoints of model construction

We initiated the chapter with the description of the two common viewpoints on subsurface property and interface detection in Section 1.2. The separation into two distinct approaches is clearly an over-simplification, but it captures the two main viewpoints on this problem and it is also, in our experience, the cause for a lot of misunderstanding between researchers trained in different fields.

Each approach in itself is open to criticism. The data-driven approach, being parsimonious, may lead to surprises and errors. This occurs in particular when geological features below the geophysical resolution exhibit a significant influence on the model forecast (e.g., Julio, Caumon, & Ford, 2015). In addition, data-driven inversions can lead to subsurface realizations that are in clear conflict with geological concepts or observations (e.g., Jessell et al., 2010). One can argue that this is not relevant if a reasonable prediction can still be obtained, but it would then mean that not all available information is considered, and also that some tacit assumptions hidden in a so-called objective model are inappropriate (Journel, 1997).

Conversely, a common criticism for the geomodel-based approach is that (a) it starts with subjective information which may induce bias in the model forecasts, (b) it contains interpretative elements which cannot be verified, and (c) it introduces a level of complexity that is not supported by data. Concerning subjectivity, we relate again to the various subjective choices that are also inherent in the data-driven approach and also state that subjectivity is not necessarily bad, as long as it is properly acknowledged and considered (Curtis, 2012). The use of all available a priori information is also strongly expressed by Tarantola (2006). Interpretative elements are certainly part of any geological investigation, and we highlight here again the hermeneutic aspect of geology (Frodeman, 1995). This is an aspect which we simply cannot avoid, and in the nature of complexity of geological investigations.

Another aspect that requires some consideration is the construction of geological models and the later usage of these models. It is a common statement that models are built for a purpose, and that this defined purpose not only is essential to define the scale of the model, but also the level of abstraction, etc. In the geoscientific literature, these approaches have recently been popularized under the general theme of fit-for-purpose models or purpose-driven geological models (Ringrose & Bentley, 2015).

In the context of geological modeling, these purpose-driven modeling approaches are typically applied in many engineering applications, for example, in reservoir modeling, but also mineral exploration and geotechnical applications. In geological modeling, however, there are many other important usages, which cannot so easily be linked to a defined purpose at the outset of the modeling study. This aspect can directly be seen when recalling that geological models are, at least partly, treated as the extension of a classical 2-D geological map into the third dimension, and it is evident in the expanding role of geological models in the context of work of geological surveys (e.g., Berg et al., 2011; Kessler et al., 2009). In this sense, geological models can be understood as an effort to extend geological knowledge, commonly and previously captured in 2-D maps, now in 3-D geological models. In these cases, it is obvious that the models are a foundation for subsequent discussions and usages which cannot be foreseen at the initial state of model construction.

Similar considerations hold in the field of hazard prediction, also a central aspect of geological surveys. Even though the model purpose is then better defined, it is not easily possible to provide a defined quantity of interest. For example, in the case of ground motion hazard predictions for Southern California (Shaw et al., 2015), a model was created to test various earthquake scenarios given ruptures at specific locations to contribute to outreach and public policies. Also in this case, the potential interest in this model will be very different for people living at different locations.

In these cases, the geological model becomes also relevant as a central communication tool. This is important, as it does not only limit the possibility to evaluate a meaningful scale and level of detail at the beginning of the modeling campaign. Also, the presentation and visualization of the model now becomes a central aspect, combined with a representation of model quality and uncertainty.

5.1.2 On the choice of a suitable geomodeling method

The various methods described in Section 3 provide ways to produce 3-D geometries from subsurface data and various degrees of conceptual knowledge.

The ability of mathematical interpolation (or approximation) to reach acceptable results has been much debated by geologists. Indeed, the ability of computers to automatically produce results acceptable by skilled structural geologists can be questionable. Early on, Walters (1969) argued that "to be effective, such a [contouring] program must allow the user to make judgments during the preparation or modification of the contoured map to enable the specialist to introduce his background". A reason for invoking expert input is that available observations never fully determine model features, which makes subsurface modeling an interpretation exercise. The interpretative aspect is, therefore, a feature to distinguish between different modeling methods. Several tools have been developed with the purpose of providing methods that allow to "draw" geological structures in 3-D, and aid in the use and combination of different data sets in this endeavor. These methods have been very successful, and widely employed, for example, in the country-wide geological model of Britain (Kessler et al., 2009). Alternatively, several methods aim at incorporating certain aspects of geology in the interpolation step itself, for example, using locally variable anisotropy (Martin & Boisvert, 2017) or prior knowledge of structural style (Caumon, Gray, et al., 2013). This also allows a significant interpretative contribution, but combines it directly with mathematical models that automatically consider certain (and perceived as important) aspects of geology. We posit that a lot of arguments for and against a certain geological modeling method are based on the perceived relevance of subjective (prior) knowledge and on how this knowledge is incorporated in the method (graphically by experts or automatically by algorithms and the associated parameters).

Overall, numerical structural modeling involves mainly two components: interpolation to fill the gaps between data points, and CAD operators to intersect, trim and weld the model elements so as to honor conceptual geological knowledge.

A difficulty at this point is to provide a comparison of the various numerical methods that does better than mentioning how "easy to use" or "flexible" they are. In practice, indeed, the level of one's familiarity with a particular software, some small but useful improvements to a particular implementation or simply different user interfaces for the same method can easily influence viewpoints. Therefore, before trying to summarize the strengths and the weaknesses of the various methods, we start by giving some evaluation criteria.

- **Expected level of geological complexity:** As already mentioned, the expressiveness of a method to complexity expected in the domain should be considered (Fig 2).
- **Computational and memory requirements:** The methods reviewed in this section do not have the same memory and computational requirements. As a rule of thumb the older methods are most efficient, whereas more recent methods are more demanding.
- **Ability to honor various data types:** Not all methods are equivalent in the type of data type they can honor.
- **Ability to incorporate geological knowledge:** The various methods all use some sort of geological knowledge, if only by letting the modeler add interpretive data in the form of maps, cross sections, or additional points. However, considering interactions between several objects (e.g., allowing for various folding styles) remains challenging with many three-dimensional methods.
- **Number of input parameters:** An objective measure of a method's simplicity is the number of parameters it involves. Note that this criterion can go against the ability to incorporate geological knowledge. In this sense, a single mathematical criterion parameterized by three or four values is preferable to one thousand mouse picks if both approaches produce a comparable result.
- **Model updating:** A geological model is seldom engraved in stone. New data may come in, new conceptual models may arise, which both require model updates. Uncertainty quantification also needs to generate not one but several possible models. It may be worth asking how many operations and how much the computational efforts are needed to update a particular model.
- **Modeling purpose:** As discussed in the Section 1.2, a geological model generally serves one or several purposes. The compatibility of a chosen modeling strategy to meet these purposes is important: the computation of a volume or the simulation of multiphase flow do not have the same implications.

With these criteria in mind, we now highlight some distinctions between explicit and implicit models, summarized in Table 1. We also refer to Jessell et al. (2014) and to Collon et al. (2016) for comparisons and discussions of the relative merits of the various numerical approaches.

Extrusion methods (depth maps and pillar approaches) are mature and have been driven by significant applications in reservoir modeling and

Table 1 Summary table comparing the various modeling approaches.

Method	**Extrusion**	**Explicit 3-D surface**	**Implicit surface**
Representativeness	Simple "2.5-D" domains	High	
Memory requirements	One 2-D scalar field	three 2D scalar fields	One 3-D scalar field
Topology	Fixed and controlled (open surface)	Fixed and controlled	Arbitrary (emerges from interpolation)
Point data	Yes, need projection		Yes, no projection involved
Orientation data	Difficult, need projection, nonlinear		Easier, no projection involved
Conformable surfaces	Iterative, involves projections		Direct
Dealing with discontinuities	Possible using Boolean operations or domain subdivision		

near-surface applications (hydrogeology and geotechnical engineering). They are conceptually simple and the associated numerical representations are compact in terms of memory and computationally efficient. They are directly compatible with prismatic gridding methods which also use surface extrusion. In particular, they can create any stratigraphic surface in the volume of interest by simple vector operations. Their main limitation is the lack of flexibility to conform to multiple geological surfaces which may be oriented along arbitrary directions. This is particularly problematic in highly deformed structural settings. Additionally, there is no direct and robust way to account for off-surface structural orientation data.

Three-dimensional explicit methods have also reached a high level of maturity. They have virtually no representational limitation in approximating geological interfaces. As they are based on two-dimensional surfaces, they also allow for a compact description of volumes. Interpolation acting on two-dimensional graphs is also relatively efficient. However, they generally involve the mastery of a relatively large set of geometrical and topological operations in order to be applied effectively. These operations (e.g., the intersection of triangulated or NURBS surfaces), although seemingly simple, are very difficult to implement in a robust way due to limited floating point accuracy. Model construction can be automated in relatively simple structural cases, but their application in complex domains generally involves significant manual editing. As data must be projected onto evolving surfaces, projections may cross discontinuities and generate local artifacts. Off-surface structural orientation data can be accounted for, assuming sedimentary beds are parallel along the projection line. Layer thickness may be locally honored using iterative projections (Salles, Ford, Joseph, Le Carlier De Veslud, & Le Solleuz, 2011), but no approach so far has automatically used cleavage or axial directions data in explicit methods. Model updating is possible but delicate if one want to maintain a valid boundary representation (Caumon et al., 2003). In terms of applications, explicit boundary representations may need to be simplified to create volumetric grids, involving stair-step approximations. Alternatively, they can create a frame for unstructured meshes, but the geostatistical modeling and process simulation are often more difficult on these meshes than on extruded regular grids.

Implicit methods are more recent, but they have raised significant enthusiasm, and have been implemented in several commercial software such as Leapfrog, Geomodeller, SKUA-GOCAD, and Petrel (Volumetric-based modeling). Several features explain this growing interest. First, implicit methods directly consider volumes and are overall more robust and automatic than surface-based models. Implicit methods leave a lot to interpolation; this means that complex shapes can be obtained automatically without need for expert input. As compared to surface-based methods, interpolation can also change the topology of the implicit surfaces. This may be a strength when the topology is unknown (e.g., for karst or ore body modeling), but may also be a problem when the topology should be controlled (for instance to prevent stratigraphic surfaces to form blobs, see Collon et al. (2016)). Another capability of implicit methods is to honor several types of structural data without any projection (Calcagno et al., 2008; Caumon, Gray, et al., 2013; Hillier et al., 2014). A continuum of conformable stratigraphic surfaces can be generated, as in extrusion approaches, but without limitations in the type of fault network. Conversely, the management of thickness variations is complex to handle in implicit methods (Laurent, 2016), as the various regularization terms used in interpolation tend to penalize simultaneously the variation of surface curvature and the variation of layer thickness. Even if recent work on anisotropy may provide interesting solutions for this (Martin & Boisvert, 2017), current implicit methods can only represent a limited number of unconformities owing to computational and storage limitations. In terms of model updating, the automation of interpolation has significantly helped the development of new structural uncertainty modeling techniques (e.g., Cherpeau et al., 2010; De la Varga et al., 2018; Lindsay et al., 2012; Wellmann et al., 2010b).

In deterministic model updates, the lack of direct surface manipulation can be a source of frustration. Indeed, the only option for local model updating is to add new data points and reinterpolate. In this case, methods that use basis functions with a local support are expected to perform more efficiently than implicit methods with a global support. As surface-based methods, implicit modeling techniques rely at some point on topological operations to finally create an image, a boundary representation or a mesh. These operations, acting on volumetric data structures, can be more difficult to implement than on surfaces.

In terms of application, a very interesting extension of implicit methods is the possibility to map every point of the subsurface into depositional space. This aspect relates not only to relative geological time but also to paleo-geographic coordinates (Mallet, 2004, 2014; Moyen et al., 2004; Wu, 2017; Wu & Hale, 2015). This representation has also been used directly for geomechanical restoration of both preserved and eroded strata (Durand-Riard, Caumon, & Muron, 2010) and it is directly compatible with the extended finite element method (Moës et al., 2002). It has also been used to create reservoir simulation grids (Gringarten et al., 2008).

5.1.3 On the classification of geological uncertainties based on uncertainty quantification concepts

In Section 2.1, we described widely used classifications for uncertainties in geological models into different types (Bardossy & Fodor, 2004; Mann, 1993). However, a difficulty in these classifications lies in the understanding of type 2 uncertainty (inherent variability). Indeed, it could be argued that there is no "inherent variability" in the natural world, as long as we stay above the scales and processes of quantum physics (e.g., leave out processes of radioactive decay).[j] Mann (1993) seems to expect a potential upscaling of Heisenberg's principle to larger scales and argues that some variables cannot be measured without altering the rock. This is, however, not similar to the uncertainty in quantum physics, as new measurement types could be able to perform these accurate measurements, and it is thus only related to a lack of knowledge. Consider the case of porosity distribution in a reservoir: we typically treat porosity as a stochastic variable and then use a random function

[j] In fact, this statement has already been made by Kitts (1976), commenting on an early work of Mann (1970).

concept to describe the spatial distribution (e.g., Chiles & Delfiner, 2012; Pyrcz & Deutsch, 2014a), and we could thus be inclined to classify porosity as a property underlying inherent variability or stochasticity, in the sense of Mann's "Type 2". However, using computer tomography (CT) methods, it is now possible to image porosity (almost) exactly (e.g., Wildenschild, Vaz, Rivers, Rikard, & Christensen, 2002)—it is thus evident that porosity is *not* underlying an inherent variability. Clearly, it is not feasible (at least with the currently available measurement techniques) to measure porosity exactly on the scale of an aquifer or a reservoir. We thus *choose* to treat porosity as a random variable, and then assign appropriate functions to model the distribution in space, for which we then can *estimate* parameters on the basis of observations.[k] We consider this example of porosity as insightful, but similar considerations would hold for the random treatment of interface positions in geological models.[l]

Another practical avenue to classify uncertainties is to consider the viewpoint of uncertainty quantification. A common approach is then to consider two classes of uncertainties (Kiureghian & Ditlevsen, 2009):

- ***Epistemic* (systematic) uncertainties:** correspond (etymologically) to uncertainties that can be described in the current state of knowledge. They cover things one could know in principle if enough information was gathered, but which cannot be observed, measured, or modeled accurately enough;
- ***Aleatory* (statistical) uncertainties:** relate the uncertainties that cannot be possibly reduced or explained other than with randomness.

As the line between aleatory and epistemic uncertainties is not very easy to draw, Kiureghian and Ditlevsen (2009) suggest to distinguish epistemic and aleatory uncertainties in the context of a specific model. Recognizing that, in principle, all uncertainties could be reduced *if* we would be able to measure exhaustively, or with a measurement device and principle which *may* be available in the future, or with more surveys that *could* in principle be made, we can consider all uncertainties as epistemic in nature. However, we separate those uncertainties that could be reduced in the near term from those that we consider as irreducible in the *current state* of a specific model. In a recent discussion on uncertainties in hydrological models, Beven (2016) classifies those uncertainties as aleatory, for which a stationary statistical

[k] For a more detailed treatment of the aspect of estimating and choosing, we refer to the insightful work of Matheron (2012).

[l] Another interesting and insightful example about the "mind projection fallacy" of randomness is described in Jaynes and Bretthorst (2003), with the common statistical example of drawing balls from an urn with replacement.

variation can be defined, and for which, therefore, the full range of statistical methods is applicable. It can be argued that this separation also adds transparency to decision-making (Kiureghian & Ditlevsen, 2009), as we clearly describe our choices about which uncertainties to consider in which way. The aim is then to further move from epistemic to aleatory residual errors, a step that could be possible with further experimentation and investigation, at least for a part of the epistemic uncertainties.

In addition to the treatment of epistemic and aleatory uncertainties, there is always the risk of "unknown unknowns", an aspect that can never be avoided. It is therefore even more important to clearly communicate the "known unknowns" that have been omitted from the analysis, for example, because of lack of knowledge, lack of computing power, or lack of time for a more detailed investigation (e.g., Beven, 2016).

In the entire discussion so far, we have left out the uncertainty related to an applied *belief system*, an aspect that can be considered as an *ontological uncertainty*, for example, the question whether probabilistic methods are suitable to represent beliefs about the nature of the residuals (e.g., Beven, 2016). This aspect is relevant, as it is related to the framework of knowledge, without which a definition of uncertainties themselves does not make sense (e.g., Nearing & Gupta, 2018), but we do not treat it in detail here and refer to the mentioned references for more information.

It remains the point that there is no single definition of types and sources of uncertainties in geological models. We are yet only at the beginning of a debate about useful definitions for subsurface models (e.g., Caers, 2011), a process that has been ongoing in related fields for many decades (e.g., in hydrology Beven, 2016; Montanari, 2007; Nearing et al., 2016). In our point of view, the consideration of aleatory and epistemic uncertainties forms an interesting framework, as this will provide a better link to concepts of uncertainty quantification in related fields. In addition, a reference to the modeling stage in which uncertainties are evaluated helps putting the work into reference (see Fig. 5). This combination will help to clarify uncertainties that have been considered in an investigation, and those that have been omitted—for reasons of deliberate choice, or because they are irreducible uncertainties in the current state (Kiureghian & Ditlevsen, 2009). It should be made absolutely clear that the sentence "How do you know that you considered *all* uncertainties?" (as common criticism to the investigation of uncertainties) does not make any sense, but it is up to the modeler to be completely clear and transparent about the uncertainties that have been considered, and those that have not. Or, as Linde, Ginsbourger, Irving, Nobile, and Doucet (2017) formulate very appropriately, "UQ in Earth Sciences can never be considered to be complete. Instead it should be viewed as a partial assessment that is valid for a given set of prior assumptions, hypotheses, and simplifications."

5.1.4 On the consideration of uncertainties in geological models

We presented in Section 4 several considerations and practical aspects for the quantification of uncertainties in 3-D geological models. The approaches mainly differ in the level of geological complexity that they allow to integrate (see Fig. 2), and the classes of uncertainties in the geological model construction step that they consider. However, in the description, we considered aspects of uncertainty that are related to modeling stages where, at least, the structure of the mathematical model has already been defined (see Fig. 5C and D). In the description above, this refers to a choice of the type of basis functions $\boldsymbol{\Phi}$, and often also the topological description $\boldsymbol{\tau}$ (i.e., the relationship between surfaces of geological interfaces and faults). But as we already pointed out in the section about geological uncertainties (Section 2), many authors stated the important relevance of uncertainties related to the initial conceptual model (Fig. 5A) and the structure of the mathematical model (Fig. 5B). These uncertainties are very difficult to capture in practice.

A first additional step is the consideration of uncertainties related to the structure of the mathematical model, for example: the number of geological interfaces and their stratigraphic and tectonic relationship, the number of faults and the topology of the fault network. The specific elements depend on the applied modeling method (see Section 3.2), but generally, these are all aspects that are related to a *single* mathematical interpolation function that has been selected for the modeling purpose. In the general model formulation of Eq. (15), this corresponds to a selection of the type of basis functions $\boldsymbol{\Phi}$, but possible differences in $\boldsymbol{\alpha}$, $\boldsymbol{\kappa}$, and $\boldsymbol{\tau}$. We will refer to differences on this level as different *model scenarios* in the following. Although several methods have been proposed to address these uncertainties using sophisticated random sampling methods (e.g., Aydin & Caers, 2017; Cherpeau & Caumon, 2015; Holden et al., 2003; Lallier et al., 2016), it is unclear at this point how they would compare to many scenarios generated by many experts.

It is evident that decisions at this stage can have an important impact on the subsequent modeling study, essentially also because these decisions are difficult to adjust in conventional geomodeling approaches. This aspect has brought several authors to argue that it is important to construct several "deterministic" models and that uncertainties in these models are more relevant than uncertainties in the stochastic model realizations (e.g., Bentley & Ringrose, 2017; Bentley & Smith, 2008; Ringrose & Bentley, 2015; Rowbotham, Kane, & Bentley, 2010).

The idea to use several model scenarios relates back to the visionary paper by Chamberlin (1897) and the concept of "multiple working hypotheses", in which he already foresees many aspects of human bias that have been formalized almost a century later by Tversky and Kahneman (1974). Following this concept, one should always formulate multiple hypotheses, in order to avoid too much "affection" (which can be understood as a combination of several bias factors) to a specific hypothesis. In the context of geological modeling, we can interpret this as an imperative to attempt to construct multiple model scenarios, each valid as a representation of the geological concept.

When we start at the level of the structure of the mathematical model, then this approach is, in fact, feasible. For example, Corbel and Wellmann (2015) described a framework for the consideration of multiple geological hypotheses when only legacy data are available, with a related high uncertainty. Fichtner et al. (2018) also recently proposed a Bayesian approach to reconcile several global or regional tomographic models created at different scales, under Gaussian assumptions. In fact, several authors also used different model scenarios as input for multiple stochastic model realizations on the basis of each of these scenarios. An interesting example is presented in the work of Suzuki et al. (2008), where then model realizations from a large set of generated models are selected on the basis of distance measures (see also Section 4.3.1) for subsequent flow simulations. Related approaches have also been taken in the field of hydrogeology, although with simpler geological modeling methods, to evaluate the effect of uncertainties in the conceptual geological model on the predictive ability of a hydrogeological model (e.g., Nilsson, Højberg, Refsgaard, & Troldborg, 2007; Refsgaard, van der Sluijs, Brown, & van der Keur, 2006; Seifert, Sonnenborg, Refsgaard, Højberg, & Troldborg, 2012; Troldborg, Refsgaard, Jensen, & Engesgaard, 2007).

One important question is how these different model scenarios are best combined with subsequent stochastic simulations in one uncertainty quantification framework. One possibility is the generation of a set of model realizations, based on different scenarios, and to apply suitable distance measures, as already mentioned above for the work of Suzuki et al. (2008), but also used widely in the field of multiple point geostatistics (e.g., Demyanov, Arnold, Rojas, & Christie, 2018; Demyanov, Backhouse, & Christie, 2015; Jung, Fenwick, & Caers, 2013; Park et al., 2013, see also Section 4.3.1).

In the context of probabilistic inverse frameworks, transdimensional MC methods provide a potentially attractive path (Sambridge et al., 2013), but they have so far only been applied to relatively simple geological model structures. An alternative would be the use of the possibility to perform structured model selection in a Bayesian framework for multiple model scenarios (see Appendix A.1). Similar approaches have been taken in hydrogeology (e.g., Schöniger, Wöhling, Samaniego, & Nowak, 2014), and an application to geomodeling can be envisaged.

Several authors have also argued that probabilistic methods are not anymore suitable when considering epistemic uncertainties related to the conceptual model (e.g., Bardossy & Fodor, 2004) and have promoted the use of extended probability methods, for example, fuzzy logic (Dubois & Prade, 2012; Zadeh, 1965). These approaches are appealing, but they have so far not been used extensively in recent uncertainty studies of geological models.

Even though it is difficult to classify all approaches, we combined the discussed methods in an overview diagram in Fig. 15. One important divide exists between the consideration of uncertainties about the conceptual model and all subsequent steps. Once the modeling method has been decided, multiple approaches exist for several purposes and levels of geological complexity.

Finally, we briefly described the link between geological models and geophysical inversions, based on several degrees of complexity in the structural models. Many developments in recent years work toward a tighter integration of information from both viewpoints, and we are confident that this will remain an active field of research, resulting in an even better combination of data and knowledge to investigate heterogeneities in the subsurface.

5.2 Conclusions and outlook

After the above review of geological modeling methods, we now aim to outline several interesting perspectives for future research in the field, as well as some recommendations, which will hopefully lead to a better integration of geological and geophysical concepts. We identify these aspects by keywords:

- **Integration:** The consideration of all available information with geological knowledge remains a serious hurdle, not only because of methodological or fundamental limitations, but also because of the different culture and value frame between geologists, geophysicists, and engineers. Even though experts are needed in each subdiscipline, teaching the value of complementary perspectives on the subsurface of the Earth

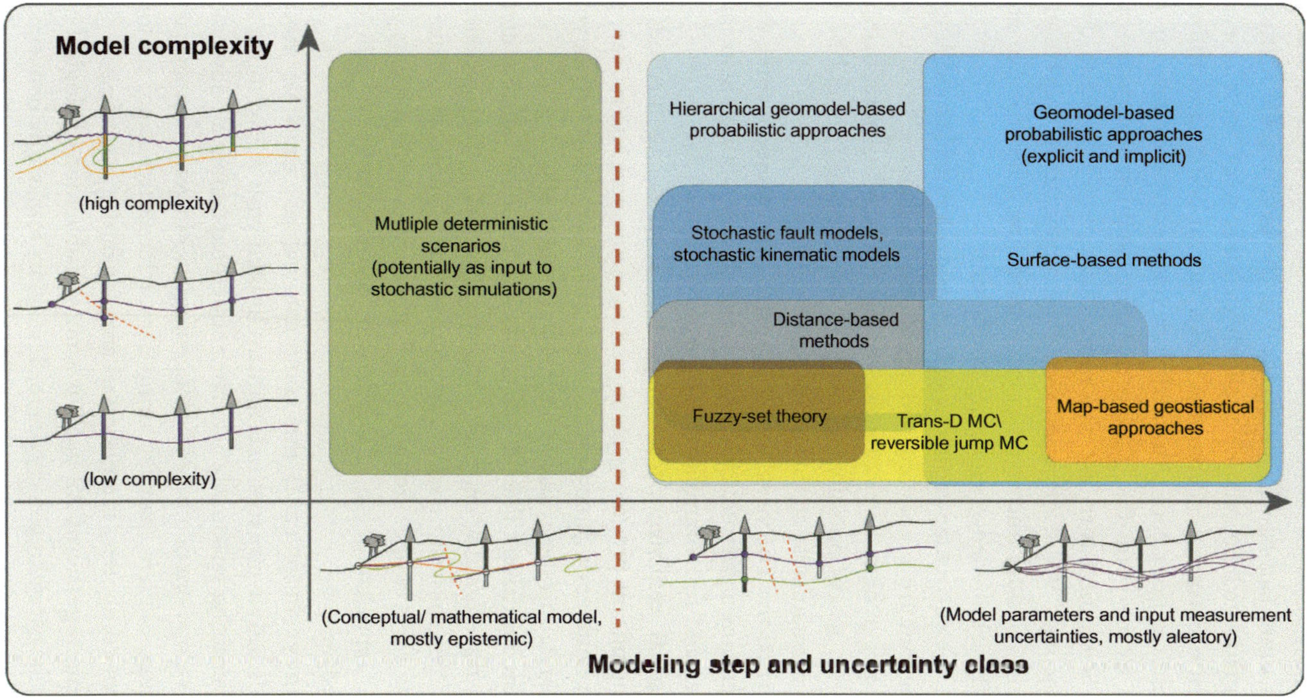

Fig. 15 Overview of methods to analyze uncertainties in structural models for different levels of geomodel complexity and different classes of uncertainties.

remains an educational challenge. Somehow, we would like to get away from strong feelings about different modeling approaches, and rather adopt a more fact-based consideration of the best possible method for a given task—which calls for explicitly defining what "best" means.

- **Communication:** Uncertainties make human beings uncomfortable; our brain (and our ego) is trained to develop convictions and to make choices to the point that choices are often implicit and not explicitly stated. Scientific papers are also more easily published when deemed "convincing". We strongly encourage geoscientists to more comprehensively consider uncertainties in all modeling endeavors, or at least to clearly acknowledge modeling assumptions.
- **Tools:** We encourage geoscientists to consider a suitable modeling *algorithm* and not just a modeling *software*. We are well aware that it is tempting to put a high level of trust in the result of a, potentially very expensive, modeling software. But the core decision should be based on the most suitable approach for a task at hand.
- **Visualization:** The visualization of spatial uncertainty in three dimensions remains a difficult and important challenge both to help making decisions and for communication purposes.
- **Geological realism:** Although geological modeling methods have made significant progresses incorporating geological knowledge, a significant number of geological concepts and observations are not yet easily integrated in the available interpolation methods.
- **Machine learning:** The advent of efficient machine learning methods opens very interesting perspectives in the field of joint geological modeling and geophysical inversion. Many interesting approaches have been available for a while and we expect interesting developments in the future, capitalizing on advances in formalizations of geological knowledge.
- **Efficiency:** In practice, the computational efficiency of modeling methods and inverse methods remains limited with regard to the ambition of creating reasonably accurate models explaining all observations. The development of efficient joint geological and geophysical inversions exploiting modern parallel computer architectures (and tomorrow, quantum computers) is an essential area for future advances.
- **Scale management:** All in all, it is not possible for a geological model to describe everything everywhere. The ability to navigate between scales depending on the considered physical processes is essential to represent geological details and uncertainties where they matter. The application

of geological modeling methods in a local way and a sound homogenization and dehomogenization are essential components for the future of joint geophysical inversion.

We want to conclude this chapter with a remark by Boschetti and Moresi (2001) on the difficulty in communication between geologists and geophysicists: "Geologists may perceive geophysicists as being lost in an abstraction far removed from real geology; geophysicists often see geologists as hopelessly resistant to mathematical rigor". With the overview presented here, we hope to contribute to a better understanding of both sides in the aim to investigate the joint object of interest: the Earth below our feet.

Acknowledgments

We are very thankful for the invitation by the editor Cédric Schmelzbach to present our viewpoint on the state of geological modeling and the related treatment of uncertainties in this issue. Comments and suggestions by Jürg Hauser and Eric de Kemp are greatly appreciated and helped significantly to clarify the content of this chapter. Furthermore, we owe great dept to the discussion about the presented topics with numerous colleagues and students over the last years and we are looking forward to many more discussions in the future! We would also like to thank the sponsors of the RING Consortium (http://www.ring.georessources.univ-lorraine.fr) for providing significant support to Guillaume Caumon and his team, and Emerson/Paradigm for providing the SKUA-GOCAD software used to prepare some figures.

Appendix

A.1 The probabilistic viewpoint on uncertainty quantification

In the chapter, we described approaches for uncertainty quantification without the direct link to a specific ontological system. However, we mentioned the use of suitable probability distributions in Section 2.2.2 and the application of these distributions in Monte Carlo simulations (Section 4.1) to estimate uncertainty propagation. Here, we briefly outline the Bayesian viewpoint for inverse approaches in the context of geological modeling.

In the Bayesian viewpoint, degrees of belief can be described using probabilities if they follow the Cox axioms (Cox, 1946). Probabilities can therefore be used to express assumptions, and equally to perform inferences given these assumptions. The important point is that these beliefs can then be communicated, as assumptions need then to be clearly expressed. Due to the rules of probability, we can then at least ensure that the same conclusions are drawn on the basis of the same assumptions (e.g., MacKay, 2003).

This viewpoint of considering beliefs as probabilities is also known as the subjective interpretation of probability (e.g., MacKay, 2003), but it has to be noted that this is neither an unwanted, nor a limiting aspect. Many authors even state that this way of thinking is essential in the context of geoscientific investigations (e.g., Curtis, 2012), and we share this viewpoint at least to the extent that we consider the aspect of geological knowlege as a central part in the context of geological modeling and geophysical inversions (see Fig. 3), and the relevance of subjectivity in the hermeneutic interpretation of geological reasoning (see Frodeman, 1995).

The concept of Bayesian statistics is described at length in many textbooks (e.g., Gelman et al., 1995; Sivia, 2006). We use here the description in MacKay (2003) using Bayes' theorem to combine the unknown parameters $\boldsymbol{\theta}$ with data D under the explicit consideration of the hypothesis space $\mathcal{H}$:

$$P(\boldsymbol{\theta}|D,\mathcal{H}) = \frac{P(D|\boldsymbol{\theta},\mathcal{H})\ P(\boldsymbol{\theta}|\mathcal{H})}{P(D|\mathcal{H})} \tag{A.1}$$

or, written in a descriptive form:

$$\text{posterior} = \frac{\text{likelihood} \times \text{prior}}{\text{evidence}} \tag{A.2}$$

For generality, we are here using $P()$ equally to denote probability densities over continuous variables and probabilities of discrete variables.

In the following description and review of concepts, we will attempt to follow this logic as closely as possible, to delineate and explain the different effects of parameters, observations, and the hypothesis space in the context of different geological modeling and uncertainty quantification processes.

In the frame of this review, we do not provide an exhaustive overview of these distributions, but refer to relevant literature. Potential prior distributions for parameters $\boldsymbol{\theta}$ as input to geological models have briefly been discussed in Section 2.2.2. Several aspects of geological likelihood functions are discussed in Section 4.3.1.

An additional point that requires consideration is that we are using a model-based Bayesian approach. In all but the simplest geological (and, equally, geophysical) modeling and inversion scenarios, a nontrivial forward model is involved. In the case of geological modeling, it can, for example, be any of the model types described in Section 3. When we consider the comparison to geophysical data, then geophysical forward simulations become an additional essential aspect. The requirements for combined geological and geophysical investigations using probabilistic methods are therefore somewhat different to many of the basic statistical models that are often used in introductory texts, as our *generative models* are generally much more complex. In addition, it is often not completely clear which role measurements and observations take (they are equally used as input parameters, but also as data in likelihood functions), and it is often impossible to even closely characterize the hypothesis space $\mathcal{H}$, which would, for example, consist of the set of all possible model scenarios.

Following MacKay (2003), we therefore also consider the possibility to compare alternative assumptions or model scenarios $\mathcal{H}$, also on the grounds of Bayes' theorem:

$$P(\mathcal{H}|D,I) = \frac{P(D|\mathcal{H},I)\ P(\mathcal{H}|I)}{P(D|I)} \tag{A.3}$$

where I denotes any additional unquestioned assumptions (e.g., continuity in space, fundamental geological and/or physical laws). In our context, different hypotheses can also be related to parameters of the mathematical model (also referred to as metaparameters or hyperparameters), see also the description in Section 2.1 and Fig. 5. The relevance of these different hypotheses has recently been highlighted for the case of seismic inversions (Thore, 2015), and we can equally consider them as highly relevant for geological modeling studies.

Furthermore, we can use the estimated uncertainty about different models to then make predictions $\boldsymbol{t}$, which are based on the observed Data D, the different model assumptions $\mathcal{H}$, and the background information I taking the probabilities of the assumptions or models into account (MacKay, 2003):

$$P(\boldsymbol{t}|D,I) = \sum_{\mathcal{H}} P(\boldsymbol{t}|D,\mathcal{H},I)\ P(\mathcal{H}|D,I) \tag{A.4}$$

For the common cases where decisions are the final aim of a geological modeling investigation, it is directly obvious that the possibility to take uncertainties about assumptions and model states into account (Eq. A.3), in combination with the adjusted parameters $\boldsymbol{\theta}$ given observed data D (Eq. A.1), can provide important insights.

A.2 Glossary

2.5-D representation of a three-dimensional model by extrusion (or sweeping) of a surface. This includes map-based representations of horizons and methods based on the extrapolation of a cross section.

Boundary representation definition of a volume by one or several water-tight surfaces.

Constraint defines a domain of nonadmissible parameter values in optimization theory; more generally, describes a piece of geological information translated in numerical terms.

Conceptual (geological) model abstract representation of geological concepts that can be understood and discussed. Often conveys temporal and causal considerations.

Data-driven approach that primarily trusts the data (and relies on a relatively simple conceptual model).

Explicit mode a *geomodel* where geological interfaces are represented as two-dimensional networks of nodes and basis functions interpolating between these nodes.

Geological interface a surface separating two geological units of different nature or age, for example: a fault, a horizon, an intrusive boundary.

Geomodel, Geological model representation of geometric elements in the subsurface, also *structural geological model.*

Geologic realism essential aspects of the complex evolution of a specific part of the Earth, considered for the purpose of a geoscientific study.

Geomodel-driven approach that uses relatively complex conceptual models which may compensate for the lack of data.

Implicit model a *geomodel* where geological interfaces are represented as level sets (or iso-value surfaces) of one or several scalar fields.

Kriging a linear unbiased interpolation method which takes a model or spatial variability (the variogram) and a model of spatial average.

Mathematical (interpolation) model a numerical framework to fill the gaps and estimate unsampled values from neighboring sample points. This involves a set of interpolation parameters $\boldsymbol{\alpha}$.

Sources of uncertainty relates to the origin of uncertainties, such as lack of knowledge, measurement error, or ambiguity.

Stochastic modeling an approach sampling several random variables to produce a set of possible models (also termed realizations).

Structural geological model representation of geometric elements in the subsurface, also *geological model.*

Topology in this paper, concerns the connectivity and adjacency between geological entities (rock units, interfaces, and lines).

Types of uncertainty relates to what could reduce the uncertainties (better measures, better data, better concepts, irreducible).

References

Abrahamsen, P. (1993). Bayesian kriging for seismic depth conversion of a multi-layer reservoir. In A. Soares (Ed.), *Geostatistics tróia '92* (pp. 385–398). Dordrecht: Springer.

Abrahamsen, P., Dahle, P., Hauge, V. L., Almendral-Vazquez, A., & Vigsnes, M. (2015). Surface prediction using rejection sampling to handle non-linear constraints. *Bulletin of Canadian Petroleum Geology*, *63*(4), 304–317.

Abrahamsen, P., Hauge, R., Heggland, K., & Mostad, P. (2000). Estimation of gross rock volume of filled geological structures with uncertainty measures. *3*(04), 304–309.

Abrahamsen, P., Lia, O., & Omre, H. (1992). *An integrated approach to prediction of hydrocarbon in place and recoverable reserve with uncertainty measures*. Society of Petroleum Engineers.

Abrahamsen, P., Omre, H., & Lia, O. (1991). *Stochastic models for seismic depth conversion of geological horizons*. Society of Petroleum Engineers.

Aitken, A., Holden, E.-J., & Dentith, M. C. (2013). Semiautomated quantification of the influence of data richness on confidence in the geologic interpretation of aeromagnetic maps. *Geophysics*, *78*(2), J1–J13.

Alcalde, J., Bond, C. E., Johnson, G., Butler, R. W. H., Cooper, M. A., & Ellis, J. F. (2017). The importance of structural model availability on seismic interpretation. *Journal of Structural Geology*, *97*, 161–171.

Alcalde, J., Bond, C. E., & Randle, C. H. (2017). Framing bias: The effect of figure presentation on seismic interpretation. *Interpretation*, *5*(4), T591–T605.

Alonso, J. L., & Teixell, A. (1992). Forelimb deformation in some natural examples of fault-propagation folds. In K. R. McClay (Ed.), *Thrust tectonics* (pp. 175–180). Dordrecht, Netherlands: Springer. https://doi.org/10.1007/978-94-011-3066-0_15http://link.springer.com/10.1007/978-94-011-3066-0_15.

Amenta, N., & Bern, M. (1999). Surface reconstruction by Voronoi filtering. *Discrete & Computational Geometry*, *22*(4), 481–504.

Anderson, M. P. (1989). Hydrogeologic facies models to delineate large-scale spatial trends in glacial and glaciofluvial sediments. *GSA Bulletin*, *101*(4), 501–511.

Armstrong, M., Galli, A., Beucher, H., Loc'h, G., Renard, D., Doligez, B., ... Geffroy, F. (2011). *Plurigaussian simulations in geosciences*. Springer Science & Business Media.

Aspinall, W. (2010). A route to more tractable expert advice. *Nature*, *463*(7279), 294–295.

Aster, R., Borchers, B., & Thurber, C. (2005). *Parameter estimation and inverse problems*. Academic Press.

Auerbach, S., & Schaeben, H. (1990). Computer-aided geometric design of geologic surfaces and bodies. *Mathematical Geology*, *22*(8), 957–987.

Aug, C., Chilès, J.-P., Courrioux, G., Lajaunie, C., Leuangthong, O., & Deutsch, C. V. (2005). 3D geological modelling and uncertainty: The potential-field method. *Geostatistics banff 2004*. In O. Leuangthong & C. V. Deutsch (Eds.), *Quantitative geology and geostatistics* (pp. 145–154). Dordrecht: Springer. http://www.springerlink.com/index/x72657121208kt67.pdf.

Autin, J., Scheck-Wenderoth, M., Götze, H. J., Reichert, C., & Marchal, D. (2016). Deep structure of the Argentine margin inferred from 3D gravity and temperature modelling, Colorado Basin. *Tectonophysics*, *676*, 198–210.

Avseth, P., Mukerji, T., & Mavko, G. (2010). *Quantitative seismic interpretation: Applying rock physics tools to reduce interpretation risk*. Cambridge University Press.

Aydin, O., & Caers, J. K. (2017). Quantifying structural uncertainty on fault networks using a marked point process within a Bayesian framework. *Tectonophysics*, *712–713*, 101–124. https://doi.org/10.1016/j.tecto.2017.04.027.

Backus, G. E., & Gilbert, J. F. (1967). Numerical applications of a formalism for geophysical inverse problems. *Geophysical Journal International*, *13*(1–3), 247–276.

Baddeley, M. C., Curtis, A., & Wood, R. (2004). An introduction to prior information derived from probabilistic judgements: Elicitation of knowledge, cognitive bias and herding. *Geological Society, London, Special Publications*, *239*(1), 15–27.

Bardossy, G., & Fodor, J. (2004). *Evaluation of uncertainties and risks in geology: New mathematical approaches for their handling*. Berlin, Germany: Springer.

Batty, M. (1974). Spatial entropy. *Geographical Analysis*, *6*(1), 1–31.

Beaumont, M. A., Zhang, W., & Balding, D. J. (2002). Approximate Bayesian computation in population genetics. *Genetics*, *162*(4), 2025–2035.

Beni, L. H., Mostafavi, M. A., & Pouliot, J. (2009). Developing an adaptive topological tessellation for 3D modeling in geosciences. *Geomatica*, *63*(4), 419–431.

Benke, K. K., Pettit, C. J., & Lowell, K. E. (2011). Visualisation of spatial uncertainty in hydrological modelling. *Journal of Spatial Science*, *56*(1), 73–88.

Bentley, M., & Ringrose, P. (2017). Future directions in reservoir modelling: New tools and 'fit-for-purpose' workflows. In *Geological Society, London, Petroleum Geology Conference Series 8* (p. p. PGC8.40).

Bentley, M., & Smith, S. (2008). Scenario-based reservoir modelling: The need for more determinism and less anchoring. *Geological Society, London, Special Publications*, *309*(1), 145–159.

Berg, R. C., Mathers, S. J., Kessler, H., & Keefer, D. A. (2011). *Synopsis of current three-dimensional geological mapping and modeling in geological survey organizations*. No. Circular 578.

Beven, K. (2016). Facets of uncertainty: Epistemic uncertainty, non-stationarity, likelihood, hypothesis testing, and communication. *Hydrological Sciences Journal*, *61*(9), 1652–1665.

Beven, K., & Binley, A. (2014). GLUE: 20years on. *Hydrological Processes*, *28*(24), 5897–5918.

Beven, K., Lamb, R., Leedal, D., & Hunter, N. (2014). Communicating uncertainty in flood inundation mapping: A case study. *International Journal of River Basin Management*, *13*(3), 285–295.

Bianchi, M., Kearsey, T., & Kingdon, A. (2015). Integrating deterministic lithostratigraphic models in stochastic realizations of subsurface heterogeneity. Impact on predictions of lithology, hydraulic heads and groundwater fluxes. *Journal of Hydrology*, *531*, 557–573.

Bilotti, F., Shaw, J. H., & Brennan, P. A. (2000). Quantitative structural analysis with stereoscopic remote sensing imagery. *AAPG Bulletin*, *84*(6), 727–740.

Bistacchi, A., Balsamo, F., Storti, F., Mozafari, M., Swennen, R., Solum, J., … Taberner, C. (2015). Photogrammetric digital outcrop reconstruction, visualization with textured surfaces, and three-dimensional structural analysis and modeling: Innovative methodologies applied to fault-related dolomitization (Vajont Limestone, Southern Alps, Italy). *Geosphere*, *11*(6), 2031–2048. https://doi.org/10.1130/GES01005.1.

Bistacchi, A., Massironi, M., Dal Piaz, V. G., Monopoli, B., Schiavo, A., & Toffolon, G. (2008). 3D fold and fault reconstruction with an uncertainty model: An example from an Alpine tunnel case study. *Computers & Geosciences*, *34*(4), 351–372.

Bobach, T., Farin, G., Hansford, D., & Umlauf, G. (2009). Natural neighbor extrapolation using ghost points. *Computer-Aided Design*, *41*(5), 350–365. https://doi.org/10.1016/j.cad.2008.08.007.

Bond, C. E., Gibbs, A. D., Shipton, Z. K., & Jones, S. (2007). What do you think this is? "Conceptual uncertainty" in geoscience interpretation. *GSA Today*, *17*(11), 4.

Bond, C. E., Johnson, G., & Ellis, J. F. (2015). Structural model creation: The impact of data type and creative space on geological reasoning and interpretation. *Geological Society, London, Special Publications*, *421*(1), SP421.4–97.

Bond, C. E., Lunn, R. J., Shipton, Z. K., & Lunn, A. D. (2012). What makes an expert effective at interpreting seismic images? *Geology*, *40*(1), 75–78.

Borgomano, J. R. F., Fournier, F., Viseur, S., & Rijkels, L. (2008). Stratigraphic well correlations for 3-D static modeling of carbonate reservoirs. *AAPG bulletin*, *92*(6), 789–824.

Bosch, M. (1999). Lithologic tomography: From plural geophysical data to lithology estimation. *Journal of Geophysical Research: Solid Earth*, *104*(B1), 749–766.

Bosch, M., Guillen, A., & Ledru, P. (2001). Lithologic tomography: An application to geophysical data from the Cadomian belt of northern Brittany, France. *Tectonophysics*, *331*(1–2), 197–227.

Bosch, M., Mukerji, T., & Gonzalez, E. F. (2010). Seismic inversion for reservoir properties combining statistical rock physics and geostatistics: A review. *Geophysics*, *75*(5), 75A165–75A176. https://doi.org/10.1190/1.3478209.

Boschetti, F., & Moresi, L. (2001). Interactive inversion in geosciences. *Geophysics*, *66*(4), 1226–1234.

Bostrom, A., Anselin, L., & Farris, J. (2008). Visualizing seismic risk and uncertainty. *Annals of the New York Academy of Sciences*, *1128*(1), 29–40.

Botter, C., Cardozo, N., Hardy, S., Lecomte, I., Paton, G., & Escalona, A. (2016). Seismic characterisation of fault damage in 3D using mechanical and seismic modelling. *Marine and Petroleum Geology*, 77, 973–990. https://doi.org/10.1016/j.marpetgeo.2016.08.002.

Bourges, M., Mari, J.-L., & Jeannée, N. (2012). A practical review of geostatistical processing applied to geophysical data: Methods and applications: Geostatistical processing applied to geophysical data. *Geophysical Prospecting*, *60*(3), 400–412. https://doi.org/10.1111/j.1365-2478.2011.00992.x.

Bowden, R. A. (2004a). Building confidence in geological models. *Geological Society, London, Special Publications*, *239*(1), 157–173.

Bowden, R. A. (2004b). Optimal elicitation of probabilistic information from experts. *Geological Society, London, Special Publications*, *239*(1), 127–145.

Brandel, S., Schneider, S., Perrin, M., Guiard, N., Rainaud, J.-F., Lienhard, P., & Bertrand, Y. (2005). Automatic building of structured geological models. *Journal of Computing and Information Science in Engineering*, *5*(2), 138. https://doi.org/10.1115/1.1884145.

Brown, R. (2004). *Animated visual vibrations as an uncertainty visualisation technique*. New York, NY: ACM.

Bruen, M., Krahe, P., Zappa, M., Olsson, J., Vehvilainen, B., Kok, K., & Daamen, K. (2010). Visualizing flood forecasting uncertainty: Some current European EPS platforms—COST731 working group 3. *Atmospheric Science Letters*, *11*(2), 92–99.

Buttenfield, B. (1993). Representing data quality. *Cartographica: The International Journal for Geographic Information and Geovisualization*, *30*(2–3), 1–7.

Caers, J. (2011). *Modeling Uncertainty in the earth sciences*. Chichester, UK: John Wiley & Sons, Ltd.

Calcagno, P., Chiles, J.-P., Courrioux, G., & Guillen, A. (2008). Geological modelling from field data and geological knowledge: Part I. Modelling method coupling 3D potential-field interpolation and geological rules: Recent advances in computational geodynamics: Theory, numerics and applications. *Physics of the Earth and Planetary Interiors*, *171*(1–4), 147–157.

Cardozo, N., & Aanonsen, S. (2009). Optimized trishear inverse modeling. *Journal of Structural Geology*, *31*(6), 546–560. https://doi.org/10.1016/j.jsg.2009.03.003.

Cardozo, N., & Brandenburg, J. P. (2014). Kinematic modeling of folding above listric propagating thrusts. *Journal of Structural Geology*, *60*, 1–12.

Carmichael, T., & Ailleres, L. (2016). Method and analysis for the upscaling of structural data. *Journal of Structural Geology*, *83*, 121–133. https://doi.org/10.1016/j.jsg.2015.09.002.
Carr, C. J., Beatson, K. R., Cherrie, B. J., Mitchell, J. T., Fright, R. W., McCallum, C. B., & Evans, R. T. (2001). Reconstruction and representation of 3D objects with radial basis functions. In *Proceedings of the 28th annual conference on computer graphics and interactive techniques* (pp. 67–76).
Catuneanu, O., Galloway, W. E., Kendall, C. G. S. C., Miall, A. D., Posamentier, H. W., Strasser, A., & Tucker, M. E. (2011). Sequence stratigraphy: Methodology and nomenclature. *Newsletters on Stratigraphy*, *44*(3), 173–245. https://doi.org/10.1127/0078-0421/2011/0011.
Caumon, G. (2010). Towards stochastic time-varying geological modeling. *Mathematical Geosciences*, *42*(5), 555–569.
Caumon, G. (2018). Geological objects and physical parameter fields in the subsurface: A review. In B. S. Daya Sagar, Q. Cheng, & F. Agterberg (Eds.), *Handbook of mathematical geosciences* (pp. 567–588). Springer International Publishing. https://doi.org/10.1007/978-3-319-78999-6_28http://link.springer.com/10.1007/978-3-319-78999-6_28.
Caumon, G., Collon-Drouaillet, P., Le Carlier De Veslud, C., Viseur, S., & Sausse, J. (2009). Surface-based 3D modeling of geological structures. *Mathematical Geosciences*, *41*(8), 927–945.
Caumon, G., Gray, G., Antoine, C., & Titeux, M. O. (2013). Three-dimensional implicit stratigraphic model building from remote sensing data on tetrahedral meshes: Theory and application to a regional model of La Popa Basin, NE Mexico. *IEEE Transactions on Geoscience and Remote Sensing*, *51*(3), 1613–1621. https://doi.org/10.1109/TGRS.2012.2207727.
Caumon, G., Laurent, G., Cherpeau, N., Lallier, F., Merland, R., Pellerin, J., & Bonneau, F. (2013). Structural framework and reservoir gridding: Current bottlenecks and way forward. In D. Garner, D. Thenin, & C. V. Deutsch (Eds.), *Closing the gap: Advances in applied geomodeling for hydrocarbon reservoirs* (pp. 43–52).
Caumon, G., Lepage, F., Sword, C. H., & Mallet, J.-L. (2004). Building and editing a sealed geological model. *Mathematical Geology*, *36*(4), 405–424.
Caumon, G., Sword Jr, C. H., & Mallet, J.-L. (2003). Constrained modifications of non-manifold B-reps. In *Proceedings of the eighth ACM symposium on solid modeling and applications* (pp. 310–315). ACM. https://doi.org/10.1145/781606.781657.
Caumon, G., Tertois, A. L., & Zhang, L. (2007). Elements for stochastic structural perturbation of stratigraphic models. *EAGE*.
Cawood, A. J., Bond, C. E., Howell, J. A., Butler, R. W. H., & Totake, Y. (2017). LiDAR, UAV or compass-clinometer? Accuracy, coverage and the effects on structural models. *Journal of Structural Geology*, *98*, 67–82.
Chamberlin, T. C. (1897). Studies for students: The method of multiple working hypotheses. *The Journal of Geology*, *5*(8), 837–848.
Charles, T., Guemene, J. M., Corre, B., Vincent, G., & Dubrule, O. (2001). *Experience with the quantification of subsurface uncertainties*. Society of Petroleum Engineers.
Cherpeau, N., & Caumon, G. (2015). Stochastic structural modelling in sparse data situations. *Petroleum Geoscience*, *21*(4), 233–247.
Cherpeau, N., Caumon, G., Caers, J., & Levy, B. (2012). Method for stochastic inverse modeling of fault geometry and connectivity using flow data. *Mathematical Geosciences*, *44*(2), 147–168.
Cherpeau, N., Caumon, G., & Lévy, B. (2010). Stochastic simulations of fault networks in 3D structural modeling. *Comptes Rendus Geoscience*, *342*(9), 687–694. https://doi.org/10.1016/j.crte.2010.04.008.
Chilès, J. P., Aug, C., Guillen, A., & Lees, T. (2004). Modelling the geometry of geological units and its uncertainty in 3D from structural data: The potential-field method. In *Proc. orebody modelling and strategic mine planning* (pp. 313–320).

Chiles, J.-P., & Delfiner, P. (2012). *Geostatistics: Modeling spatial uncertainty*. In *Wiley series in probability and statistics*. Hoboken, NJ: Wiley.
Clapp, R. G., Biondi, B. L., & Claerbout, J. F. (2004). Incorporating geologic information into reflection tomography. *Geophysics*, *69*(2), 533–546. https://doi.org/10.1190/1.1707073.
Collon, P., Pichat, A., Kergaravat, C., Botella, A., Caumon, G., Ringenbach, J.-C., & Callot, J.-P. (2016). 3D modeling from outcrop data in a salt tectonic context: Example from the Inceyol minibasin, Sivas Basin, Turkey. *Interpretation*, *4*(3), SM17–SM31. https://doi.org/10.1190/INT-2015-0178.1.
Collon-Drouaillet, P., Steckiewicz-Laurent, W., Pellerin, J., Laurent, G. Caumon, G., Reichart, G., & Vaute, L. (2015). 3D geomodelling combining implicit surfaces and voronoi-based remeshing: A case study in the Lorraine Coal Basin (France). *Computers & Geosciences*, *77*, 29–43. https://doi.org/10.1016/j.cageo.2015.01.009.
Comber, A., Fisher, P., Brunsdon, C., & Khmag, A. (2012). Spatial analysis of remote sensing image classification accuracy. *Remote Sensing of Environment*, *127*, 237–246.
Cooke, R. M. (1991). *Experts in uncertainty: Opinion and subjective probability in science*. Oxford University Press.
Corbel, S., & Wellmann, F. (2015). Framework for multiple hypothesis testing improves the use of legacy data in structural geological modeling. *GeoResJ*, *6*, 202–212.
Corre, B., Thore, P., de Feraudy, V., & Vincent, G. (2000). *Integrated uncertainty assessment for project evaluation and risk analysis*. Society of Petroleum Engineers.
Courrioux, G., Nullans, S., Guillen, A., Boissonnat, J. D., Repusseau, P., Renaud, X., & Thibaut, M. (2001). 3D volumetric modelling of Cadomian terranes (Northern Brittany, France): An automatic method using Vorono diagrams. *Tectonophysics*, *331*(1–2), 181–196.
Cover, T. M., & Thomas, J. A. (2005). *Elements of information theory*. Hoboken, NJ: John Wiley & Sons, Inc.
Cowan, E. J., Beatson, R. K., Fright, W. R., McLennan, T. J., & Mitchell, T. J. (2002). Rapid geological modelling. *Applied Structural Geology for Mineral Exploration and Mining*.
Cox, A. L. (1982). Artifactual uncertainty in risk analysis. *Risk Analysis*, *2*(3), 121–135.
Cox, J., House, D., & Lindell, M. (2013). Visualizing uncertainty in predicted hurricane tracks. *International Journal for Uncertainty Quantification*, *3*(2), 143–156.
Cox, R. T. (1946). Probability, frequency and reasonable expectation. *American Journal of Physics*, *14*(1), 1–13.
Cracknell, M. J., & Reading, A. M. (2013). The upside of uncertainty: Identification of lithology contact zones from airborne geophysics and satellite data using random forests and support vector machines. *Geophysics*, *78*(3), WB113–WB126.
Cracknell, M. J., & Reading, A. M. (2014). Geological mapping using remote sensing data: A comparison of five machine learning algorithms, their response to variations in the spatial distribution of training data and the use of explicit spatial information. *Computers and Geosciences*, *63*, 22–33.
Cressie, N. (1990). The origins of kriging. *Mathematical Geology*, *22*(3), 239–252.
Culshaw, M. G. (2005). From concept towards reality: Developing the attributed 3D geological model of the shallow subsurface. *Quarterly Journal of Engineering Geology and Hydrogeology*, *38*(3), 231–284.
Curtis, A. (2012). The science of subjectivity. *Geology*, *40*(1), 95–96.
Curtis, A., & Lomax, A. (2001). Prior information, sampling distributions, and the curse of dimensionality. *Geophysics*, *66*(2), 372–378.
Curtis, A., & Wood, R. (2004). Optimal elicitation of probabilistic information from experts. *Geological Society, London, Special Publications*, *239*(1), 127–145.
Davies, R. J., Stewart, S. A., Cartwright, J. A., Lappin, M., Johnston, R., Fraser, S. I., & Brown, A. R. (2004). 3D seismic technology: Are we realising its full potential? *Geological Society, London, Memoirs*, *29*(1), 1–10.

Davis, C. J. (2002). *Statistics and data analysis in geology*. New York, NY: Wiley.
de Kemp, E. A., & Sprague, K. B. (2003). Interpretive tools for 3-D structural geological modeling part I: Bézier-based curves, ribbons and grip frames. *Geoinformatica*, 7(1), 55–71.
De la Varga, M., Schaaf, A., & Wellmann, F. (2018). GemPy 1.0: Open-source stochastic geological modeling and inversion. *Geoscientific Model Development Discussions*, 1–50. https://doi.org/10.5194/gmd-2018-61 (in review).
De la Varga, M., Wellmann, F. (2016). Structural geologic modeling as an inference problem: A Bayesian perspective. *Interpretation*, *4*(3), 1–16.
Delfiner, P., & Chiles, J. P. (1977). *Conditional simulations: A new monte carlo approach to probabilistic evaluation of hydrocarbon in place* (p. 32). Society of Petroleum Engineers. SPE-6985-MS.
Demyanov, V., Arnold, D., Rojas, T., & Christie, M. (2018). Uncertainty quantification in reservoir prediction: Part 2—Handling uncertainty in the geological scenario. *Mathematical Geosciences*, *50*(17), 1–24.
Demyanov, V., Backhouse, L., & Christie, M. (2015). Geological feature selection in reservoir modelling and history matching with Multiple Kernel Learning. *Computers and Geosciences*, *85*, 16–25.
Dentith, M., & Mudge, S. T. (2014). *Geophysics for the mineral exploration geoscientist*. Cambridge University Press.
De Paor, D. G. (1996). B´ezier curves and geological design. In D. G. De Paor (Ed.), *Computer methods in the geosciences* (pp. 389–417). Pergamon.
Djurcilov, S., Kim, K., Lermusiaux, P., & Pang, A. (2002). Visualizing scalar volumetric data with uncertainty. *Computers & Graphics*, *26*(2), 239–248.
Djurcilov, S., Kim, K., & Lermusiaux, P. F. J., & Pang, A. (2001). Volume rendering data with uncertainty information. In *Data visualization 2001* (pp. 243–252). Vienna: Springer.
Dorn, G. A. (1998). Modern 3-D seismic interpretation. *The Leading Edge*, *17*(9), 12621262.
Dubois, D., & Prade, H. (2012). Fundamentals of fuzzy sets. In D. Dubois & H. Prade (Eds.), *The handbooks of fuzzy sets*. Springer.
Dubrule, O. (1984). Comparing splines and kriging. *Computers & Geosciences*, *10*(2–3), 327–338. https://doi.org/10.1016/0098-3004(84)90030-X.
Durand-Riard, P., Caumon, G., & Muron, P. (2010). Balanced restoration of geological volumes with relaxed meshing constraints. *Computers & Geosciences*, *36*(4), 441–452.
Edwards, J., Lallier, F., Caumon, G., & Carpentier, C. (2018). Uncertainty management in stratigraphic well correlation and stratigraphic architectures: A training-based method. *Computers & Geosciences*, *111*, 1–17. https://doi.org/10.1016/j.cageo.2017.10.008.
Eidsvik, J., Mukerji, T., & Switzer, P. (2004). Estimation of geological attributes from a well log: An application of Hidden Markov chains. *Mathematical Geology*, *36*(3), 379–397.
Elfeki, A. M. M., & Dekking, F. M. (2005). Modelling subsurface heterogeneity by coupled Markov chains: Directional dependency, walther's law and entropy. *Geotechnical & Geological Engineering*, *23*(6), 721–756.
Farmer, C. L. (2005). Geological modelling and reservoir simulation. In A. Iske & T. Randen (Eds.), *Mathematical Methods and Modelling in Hydrocarbon Exploration and Production* (pp. 119–212). Berlin, Heidelberg: Springer. https://doi.org/10.1007/3-540-26493-0_6.
Fernández, O., Muñoz, J. A., Arbués, P., Falivene, O., & Marzo, M. (2004). Three-dimensional reconstruction of geological surfaces: An example of growth strata and turbidite systems from the Ainsa basin (Pyrenees, Spain). *AAPG Bulletin*, *88*(8), 1049–1068.
Ferrill, D. A., Morris, A. P., Stamatakos, J. A., & Sims, D. W. (2000). Crossing conjugate normal faults. *AAPG Bulletin*, *84*(10), 1543–1559.
Fichtner, A., van Herwaarden, D.-P., Afanasiev, M., Simutė, S., Krischer, L., Çubuk-Sabuncu, Y., … Igel, H. (2018). The collaborative seismic earth model: Generation 1. *Geophysical Research Letters*, *45*(9), 4007–4016. https://doi.org/10.1029/2018GL077338.

Fisher, N. I., Embelton, B. J. J., & Lewis, T. (1993). *Statistical analysis of spherical data.* Cambridge, New York, NY: Cambridge University Press.

Fisher, P., Comber, A., & Wadsworth, R. (2006). Approaches to uncertainty in spatial data. In *Fundamentals of spatial data quality* (pp. 43–59). London, UK: ISTE.

Fisher, R. (1953). Dispersion on a sphere. *Proceedings of the royal society of London. Series A, mathematical and physical sciences, 217*(1130), 295.

Fisher, T. R., & Wales, R. Q. (1992). Three dimensional solid modeling of geo-objects using non-uniform rational B-splines (NURBS). In A. K. Turner (Ed.), *Three-dimensional modeling with geoscientific information systems* (pp. 85–105). Dordrecht, Netherlands: Springer.

Ford, A., & Blenkinsop, T. G. (2008). Evaluating geological complexity and complexity gradients as controls on copper mineralisation, Mt Isa Inlier. *Australian Journal of Earth Sciences, 55*(1), 13–23.

Ford, D. N., & Sterman, J. D. (1998). Expert knowledge elicitation to improve formal and mental models. *System Dynamics Review, 14*(4), 309–340.

Frank, T., Tertois, A.-L., & Mallet, J.-L. (2007). 3D-reconstruction of complex geological interfaces from irregularly distributed and noisy point data. *Computers & Geosciences, 33*(7), 932–943. https://doi.org/10.1016/j.cageo.2006.11.014.

Fremming, N. P. (2002). 3D geological model construction using a 3D grid. In *ECMOR VIII-8th European conference on the mathematics of oil recovery.*

Frodeman, R. (1995). Geological reasoning: Geology as an interpretive and historical science. *GSA Bulletin, 107*(8), 960–968.

Fullagar, P. K., Pears, G. A., & McMonnies, B. (2008). Constrained inversion of geologic surfaces—Pushing the boundaries. *The Leading Edge, 27*(1), 98–105.

Gallardo, L. A., & Meju, M. A. (2003). Characterization of heterogeneous near-surface materials by joint 2D inversion of dc resistivity and seismic data. *Geophysical Research Letters, 30*(13), 3018.

Gallardo, L. A., & Meju, M. A. (2011). Structure-coupled multiphysics imaging in geophysical sciences. *Reviews of Geophysics, 49*(1), 3115.

Galli, A., Beucher, H., Le Loc'h, G., & Doligez, B. (1994). The pros and cons of the truncated Gaussian method. In M. Armstrong & P. A. Dowd (Eds.), *Geostatistical simulations* (pp. 217–233). Springer.

Gawith, D. E., & Gutteridge, P. A. (1996). Seismic validation of reservoir simulation using a shared earth model. *Petroleum Geoscience, 2*(2), 97–103.

Gelman, A., Carlin, J. B., Stern, H. S., & Rubin, D. B. (1995). *Bayesian data analysis.* Chapman and Hall/CRC.

Georgsen, F., Røe, P., Syversveen, A. R., & Lia, O. (2012). Fault displacement modelling using 3D vector fields. *Computational Geosciences, 16*(2), 247–259. https://doi.org/10.1007/s10596-011-9257-z.

Giles, K. A., & Lawton, T. F. (2002). Halokinetic sequence stratigraphy adjacent to the El Papalote diapir, Northeastern Mexico. *AAPG Bulletin, 86*(5), 823–840. https://doi.org/10.1306/61EEDBAC-173E-11D7-8645000102C1865D.

Giraud, J., Pakyuz-Charrier, E., Jessell, M. W., Lindsay, M., Martin, R., & Ogarko, V. (2017). Uncertainty reduction through geologically conditioned petrophysical constraints in joint inversionConditioned petrophysical constraints. *Geophysics, 82*(6), ID19–ID34.

Gjøystdal, H., Reinhardsen, J. E., & Åstebøl, K. (1985). Computer representation of complex 3-D geological structures using a new "sold modeling" technique. *Geophysical Prospecting, 33*(8), 1195–1211.

Glinsky, E. M., Asher, B., Hill, R., Flynn, M., Stanley, M., Gunning, J., … White, C. (2005). Integration of uncertain subsurface information into multiple reservoir simulation models. *The Leading Edge, 24*(10), 990–999.

Goda, K., & Song, J. (2015). Uncertainty modeling and visualization for tsunami hazard and risk mapping: A case study for the 2011 Tohoku earthquake. *Stochastic Environmental Research and Risk Assessment, 30*(8), 2271–2285.

Godefroy, G. (2018). Modlisation cinmatique et stochastique des failles partir de donnes parses pour lanalyse des incertitudes structurales (Unpublished doctoral dissertation). Universit de Lorraine.

Godefroy, G., Caumon, G., Ford, M., Laurent, G., & Jackson, C. A. L. (2018). A parametric fault displacement model to introduce kinematic control into modeling faults from sparse data. *Interpretation, 6*(2), 1–14. https://doi.org/10.1190/int-2017-0059.1.

Gonçalves, Í. G., Kumaira, S., & Guadagnin, F. (2017). A machine learning approach to the potential-field method for implicit modeling of geological structures. *Computers and Geosciences, 103*, 173–182.

Goodchild, F. M., Buttenfield, B., & Wood, J. (1994). Introduction to visualizing data validity. In M. H. Hearnshaw & J. D. Unwin (Eds.), *Visualization in geographical information systems* (pp. 141–149). New York, NY: John Wiley & Sons.

Goodchild, F. M., Chih-Chang, L., & Leung, Y. (1994). Visualizing fuzzy maps. In M. H. Hearnshaw & J. D. Unwin (Eds.), *Visualization in geographical information systems* (pp. 158–167). New York, NY: John Wiley & Sons.

Goovaerts, P. (1997). *Geostatistics for natural resources evaluation.* Oxford University Press.

Götze, H. J., & Lahmeyer, B. (1988). Application of three-dimensional interactive modeling in gravity and magnetics. *Geophysics, 53*(8), 1096–1108.

Gradmann, S., Ebbing, J., & Fullea, J. (2013). Integrated geophysical modelling of a lateral transition zone in the lithospheric mantle under Norway and Sweden. *Geophysical Journal International, 194*(3), 1358–1373.

Grana, D. (2018). Joint facies and reservoir properties inversionJoiFRI. *Geophysics, 83*(3), M15–M24.

Grana, D., Pirrone, M., & Mukerji, T. (2012). Quantitative log interpretation and uncertainty propagation of petrophysical properties and facies classification from rock-physics modeling and formation evaluation analysis. *Geophysics*, 77(3), WA45–WA63.

Gratier, J.-P., & Guillier, B. (1993). Compatibility constraints on folded and faulted strata and calculation of total displacement using computational restoration (UNFOLD program). *Journal of Structural Geology, 15*(3–5), 391–402. https://doi.org/10.1016/0191-8141(93)90135-W.

Greengard, L., & Rokhlin, V. (1987). A fast algorithm for particle simulations. *Journal of Computational Physics, 73*(2), 325–348. https://doi.org/10.1016/0021-9991(87)90140-9.

Greenly, E. (1930). *Methods in geological surveying.* New York: Thomas Murby And Co.

Gringarten, E. J., Arpat, G. B., Haouesse, M. A., Dutranois, A., Deny, L., Jayr, S., & Nghiem, L. X. (2008). New grids for robust reservoir modeling. *2008 SPE ATCE.* Denver: Society of Petroleum Engineers. https://doi.org/10.2118/116649-MS.

Grose, L., Laurent, G., Ailleres, L., Armit, R., Jessell, M. W., & Caumon, G. (2017). Structural data constraints for implicit modeling of folds. *Journal of Structural Geology, 104*, 80–92.

Grose, L., Laurent, G., Ailleres, L., Armit, R., Jessell, M. W., & Dechenaud, T. C. (2018). Inversion of structural geology data for fold geometry. *Journal of Geophysical Research: Solid Earth, 123*, 6318–6333.

Guillen, A., Calcagno, P., Courrioux, G., Joly, A., & Ledru, P. (2008). Geological modelling from field data and geological knowledge: Part II. Modelling validation using gravity and magnetic data inversion: Recent advances in computational geodynamics: Theory, numerics and applications. *Physics of the Earth and Planetary Interiors, 171*(1–4), 158–169.

Guiziou, J. L., Mallet, J. L., & Madariaga, R. (1996). 3-D seismic reflection tomography on top of the GOCAD depth modeler. *Geophysics, 61*(5), 1499–1510.

Haase, C., Ebbing, J., & Funck, T. (2017). A 3D regional crustal model of the NE Atlantic based on seismic and gravity data. *Geological Society, London, Special Publications, 447*(1), 233–247.

Hale, D., & Emanuel, J. (2003). *Seismic interpretation using global image segmentation* (pp. 2410–2413). Society of Exploration Geophysicists. https://doi.org/10.1190/1 1817860.

Hardy, R. L. (1971). Multiquadric equations of topography and other irregular surfaces. *Journal of Geophysical Research*, *76*(8), 1905–1915.

Harris, J. R. (2007). Remote predictive mapping (RPM): A not so new paradigm for mapping Canada's North. *Geoscience Canada*, *34*(3 and 4), 91–92.

Harris, J. R., Wickert, L., Lynds, T., Behnia, P., Rainbird, R., Grunsky, E., … Schetselaar, E. (2011). Remote predictive mapping 3. Optical remote sensing—A review for remote predictive geological mapping in northern canada. *Geoscience Canada*, *38*(2), 49–83.

Hauser, J., Dyer, K. M., Pasyanos, M. E., Bungum, H., Faleide, J. I., Clark, S. A., & Schweitzer, J. (2011). A probabilistic seismic model for the European Arctic. *Journal of Geophysical Research*, *116*(B1), B01303.

Hauser, J., Gunning, J., & Annetts, D. (2015). Probabilistic inversion of airborne electromagnetic data under spatial constraints. *Geophysics*, *80*(2), E135–E146.

Hauser, J., Gunning, J., & Annetts, D. (2016). Probabilistic inversion of airborne electromagnetic data for basement conductors. *Geophysics*, *81*(5), E389–E400.

Henrion, V., Caumon, G., & Cherpeau, N. (2010). ODSIM: An object-distance simulation method for conditioning complex natural structures. *Mathematical Geosciences*, *42*(8), 911–924.

Herron, D. A. (2015). Pitfalls in horizon autopicking. *Interpretation*, *3*(1), SB1–SB4.

Hillier, M. J., Schetselaar, E. M., de Kemp, E. A., & Perron, G. (2014). Three-dimensional modelling of geological surfaces using generalized interpolation with radial basis functions. *Mathematical Geosciences*, *46*(8), 931–953. https://doi.org/10.1007/s11004-014-9540-3.

Hoffman, K. S., & Neave, J. W. (2007a). The fused fault block approach to fault network modelling. *Geological Society, London, Special Publications*, *292*(1), 75–87. https://doi.org/10.1144/SP292.4.

Hoffman, K. S., & Neave, J. W. (2007b). The fused fault block approach to fault network modelling. *Geological Society, London, Special Publications*, *292*(1), 75–87.

Holden, E.-J., Wong, J. C., Kovesi, P., Wedge, D., Dentith, M., & Bagas, L. (2012). Identifying structural complexity in aeromagnetic data: An image analysis approach to greenfields gold exploration. *Ore Geology Reviews*, *46*, 47–59.

Holden, L., Mostad, P., Nielsen, B., Gjerde, J., Townsend, C., & Ottesen, S. (2003). Stochastic structural modeling. *Mathematical Geology*, *35*(8), 899–914.

Hollund, K., Mostad, P., Nielsen, B. F., Holden, L., Gjerde, J., Contursi, M. G., … Sverdrup, E. (2002). Havana—A fault modeling tool. *Norwegian Petroleum Society Special Publications*, *11*, 157–171.

Hornby, P., Boschetti, F., & Horowitz, F. (1999). Analysis of potential field data in the wavelet domain. *Geophysical Journal of the Royal Astronomical Society*, *137*(1), 175–196.

Houlding, S. W. (1994). *3D geoscience modeling computer techniques for geological characterization*. Springer-Verlag.

Hu, L. Y., & Chugunova, T. (2008). Multiple-point geostatistics for modeling subsurface heterogeneity: A comprehensive review: Review of Multiple POI. *Water Resources Research*, *44*(11), W11413. https://doi.org/10.1029/2008WR006993.

Hudec, M. R., & Jackson, M. P. A. (2007). Terra infirma: Understanding salt tectonics. *Earth-Science Reviews*, *82*(1–2), 1–28. https://doi.org/10.1016/j.earscirev.2007.01.001.

Irakarama, M., Cupillard, P., Caumon, G., & Sava, P. (2017). Appraising structural models using seismic data: Problem and challenges. *SEG technical program expanded abstracts 2017* (pp. 1897–1901). Society of Exploration Geophysicists. http://library.seg.org/doi/abs/10.1190/segam2017-17791860.1.

Irving, A. D., Chavanne, E., Faure, V., Buffet, P., & Barber, E. (2010). An uncertainty modelling workflow for structurally compartmentalized reservoirs. *Geological Society, London, Special Publications, 347*(1), 283–299.

Jackson, C. A. L., & Kane, K. E. (2012). *Recent Advances: Vol. 96. 3D seismic interpretation techniques: Applications to Basin analysis.* Chichester, UK: Wiley-Blackwell.

Jacoby, W. J., & Smilde, P. L. (2009). *Gravity interpretation.* Berlin, Heidelberg: Springer.

Jaynes, E. T., & Bretthorst, G. L. (2003). *Probability theory: The logic of science.* Cambridge University Press.

Jessell, M. W. (1981). "Noddy"—An interactive Map creation Package (Unpublished doctoral dissertation). Imperial College of Science and Technology, London.

Jessell, M. W., Ailleres, L., de Kemp, E., Lindsay, M., Wellmann, F., Hillier, M., … Martin, R. (2014). Next generation three-dimensional geologic modeling and inversion. *Society of Economic Geologists Special Publication, 18,* 261–272.

Jessell, M. W., Ailleres, L., & Kemp, A. E. (2010). Towards an integrated inversion of geoscientific data: What price of geology? *Tectonophysics, 490*(3–4), 294–306.

Jessell, M. W., & Valenta, R. K. (1996). Structural geophysics: Integrated structural and geophysical modelling. *Computer Methods in the Geosciences, 15,* 303–324.

Johnson, C. R., & Jones, T. A. (1988). Putting geology into reservoir simulations: A three-dimensional modeling approach. *SPE-18321-MS*SPE: Society of Petroleum Engineers. https://doi.org/10.2118/18321-MS.

Joly, A., Chen, Y., Faure, M., & Martelet, G. (2007). A multidisciplinary study of a syntectonic pluton close to a major lithospheric-scale fault-Relationships between the Montmarault granitic massif and the Sillon Houiller Fault in the Variscan French Massif Central: 1. Geochronology, mineral fabrics, and tectonic implications. *Journal of Geophysical Research, 112,* B10104.

Joly, A., Faure, M., Martelet, G., & Chen, Y. (2009). Gravity inversion, AMS and geochronological investigations of syntectonic granitic plutons in the southern part of the Variscan French Massif Central. *Journal of Structural Geology, 31*(4), 421–443.

Joly, A., Martelet, G., Chen, Y., & Faure, M. (2008). A multidisciplinary study of a syntectonic pluton close to a major lithospheric-scale fault—Relationships between the Montmarault granitic massif and the Sillon Houiller Fault in the Variscan French Massif Central: 2. Gravity, aeromagnetic investigations, and 3-D geologic modeling. *Journal of Geophysical Research, 113,* B01404.

Jones, R. R., McCaffrey, J. W. K., Wilson, W. R., & Holdsworth, E. R. (2004). Digital field data acquisition: Towards increased quantification of uncertainty during geological mapping. *Geological Society, London, Special Publications, 239*(1), 43.

Journel, A. G. (1997). The abuse of principles in model building and the quest for objectivity. In E. Y. Baafi & N. A. Schofield (Eds.), *Quantitative geology and geostatistics: Vol 1. Geostatistics Wollongong '96* (pp. 3–14). Kluwer Academic Publisher.

Journel, A. G., & Huijbregts, C. J. (1978). *Mining geostatistics.* Academic Press.

Julio, C., Caumon, G., & Ford, M. (2015). Impact of the en echelon fault connectivity on reservoir flow simulations. *Interpretation, 3*(4), SAC23–SAC34. https://doi.org/10.1190/INT-2015-0060.1.

Jung, A., Fenwick, D. H., & Caers, J. (2013). Training image-based scenario modeling of fractured reservoirs for flow uncertainty quantification. *Comput Geosci, 17*(6), 1015–1031.

Kabanikhin, S. I. (2008). Definitions and examples of inverse and ill-posed problems. *Journal of Inverse and Ill-posed Problems, 16*(4), 317–357.

Kamei, R., & Lumley, D. (2017). Full waveform inversion of repeating seismic events to estimate time-lapse velocity changes. *Geophysical Journal International, 209*(2), 1239–1264.

Kaufmann, O., & Martin, T. (2008). 3D Geological modelling from boreholes, cross-sections and geological maps, application over former natural gas storages in coal mines. *Computers & Geosciences, 34*(3), 278–290.

Kelk, B. (1992). 3-D modelling with geoscientific information systems: The problem. In A. K. Turner (Ed.), *Three-dimensional modeling with geoscientific information systems* (pp. 29–37). Dordrecht, Netherlands: Springer.
Kessler, H., Mathers, S., & Sobisch, H.-G. (2009). The capture and dissemination of integrated 3D geospatial knowledge at the British geological survey using GSI3D software and methodology. *Computers & Geosciences*, *35*(6), 1311–1321.
Kidd, A. L. (1987). Knowledge acquisition. *Knowledge acquisition for expert systems* (pp. 1–16). Boston, MA: Springer.
Kim, Y.-S., & Sanderson, D. J. (2005). The relationship between displacement and length of faults: A review. *Earth-Science Reviews*, *68*(3–4), 317–334. https://doi org/10.1016/j.earscirev.2004.06.003.
Kitts, D. B. (1976). Certainty and uncertainty in geology. *American Journal of Science*, *276*(1), 29–46.
Kiureghian, A. D., & Ditlevsen, O. (2009). Aleatory or epistemic? Does it matter? *Structural Safety*, *31*(2), 105–112.
Kunz, M., Grêt-Regamey, A., & Hurni, L. (2011). Visualization of uncertainty in natural hazards assessments using an interactive cartographic information system. *Natural Hazards*, *59*(3), 1735–1751.
Lajaunie, C., Courrioux, G., & Manuel, L. (1997). Foliation fields and 3D cartography in geology: Principles of a method based on potential interpolation. *Mathematical Geology*, *29*(4), 571–584.
Lallier, F., Caumon, G., Borgomano, J., Viseur, S., Fournier, F., Antoine, C., & Gentilhomme, T. (2012). Relevance of the stochastic stratigraphic well correlation approach for the study of complex carbonate settings: Application to the Malampaya buildup (Offshore Palawan, Philippines). *Geological Society, London, Special Publications*, *370*(1), 265–275.
Lallier, F., Caumon, G., Borgomano, J., Viseur, S., Royer, J.-J., & Antoine, C. (2016). Uncertainty assessment in the stratigraphic well correlation of a carbonate ramp: Method and application to the Beausset Basin, SE France. *Comptes Rendus Geoscience*, *348*(7), 499–509. https://doi.org/10.1016/j.crte.2015.10.002.
Lary, D. J., Alavi, A. H., Gandomi, A. H., & Walker, A. L. (2016). Machine learning in geosciences and remote sensing. *Geoscience Frontiers*, 7(1), 3–10.
Laurent, G. (2016). Iterative thickness regularization of stratigraphic layers in discrete implicit modeling. *Mathematical Geosciences*, *48*, 811–833. https://doi.org/10.1007/s11004-016-9637-y.
Laurent, G., Ailleres, L., Grose, L., Caumon, G., Jessell, M. W., & Armit, R. (2016). Implicit modeling of folds and overprinting deformation. *Earth and Planetary Science Letters*, *456*, 26–38.
Laurent, G., Caumon, G., Bouziat, A., & Jessell, M. (2013). A parametric method to model 3D displacements around faults with volumetric vector fields. *Tectonophysics*, *590*, 83–93. https://doi.org/10.1016/j.tecto.2013.01.015.
Laurent, G., Caumon, G., & Jessell, M. (2015). Interactive editing of 3D geological structures and tectonic history sketching via a rigid element method. *Computers & Geosciences*, *74*, 71–86.
Lecour, M., Cognot, R., Duvinage, I., Thore, P., & Dulac, J.-C. (2001). Modelling of stochastic faults and fault networks in a structural uncertainty study. *Petroleum Geoscience*, 7, 31–42.
Lelièvre, P., Oldenburg, D., & Williams, N. (2012). Constraining geophysical inversions with geologic information. *Seg technical program expanded abstracts 2008* (pp. 1223–1227). Society of Exploration Geophysicists.
Lelièvre, P. G., & Oldenburg, D. W. (2009). A comprehensive study of including structural orientation information in geophysical inversions. *Geophysical Journal International*, *178*(2), 623–637.

Lelliott, M. R., Cave, M. R., & Wealthall, G. P. (2009). A structured approach to the measurement of uncertainty in 3D geological models. *Quarterly Journal of Engineering Geology and Hydrogeology*, *42*(1), 95–105.

Lemon, A. M., & Jones, N. L. (2003). Building solid models from boreholes and user-defined cross-sections. *Computers & Geosciences*, *29*(5), 547–555.

Leung, Y., Goodchild, F. M., & Lin, C. C. (1993). Visualization of fuzzy scenes and probability fields. In H. J. Newton (Ed.), *Computing science and statistics, volume 24: Graphics and visualization* (pp. 416–422). *Proceedings of the 24th symposium on the Interface*. Fairfax Station, VA: Interface Foundation of North America.

Li, L., Caers, J., & Sava, P. (2015). Assessing seismic uncertainty via geostatistical velocity-model perturbation and image registration: An application to subsalt imaging. *The Leading Edge*, *34*(9), 1064–1070.

Li, Y., & Oldenburg, D. W. (1998). 3-D inversion of gravity data. *Geophysics*, *63*(1), 109–119.

Li, Y., & Oldenburg, D. W. (2000). Incorporating geological dip information into geophysical inversions. *Geophysics*, *65*(1), 148–157.

Lim, N. J., Brandt, S. A., & Seipel, S. (2015). Visualisation and evaluation of flood uncertainties based on ensemble modelling. *International Journal of Geographical Information Science*, *30*(2), 240–262.

Linde, N., Ginsbourger, D., Irving, J., Nobile, F., & Doucet, A. (2017). On uncertainty quantification in hydrogeology and hydrogeophysics. *Advances in Water Resources*, *110*, 166–181.

Lindsay, M., Ailleres, L., Jessell, M. W., de Kemp, E., & Betts, P. G. (2012). Locating and quantifying geological uncertainty in three-dimensional models: Analysis of the Gippsland Basin, Southeastern Australia. *Tectonophysics*, *546-547*, 10–27.

Lindsay, M., Perrouty, S., Jessell, M. W., & Ailleres, L. (2014). Inversion and geodiversity: Searching model space for the answers. *Mathematical Geosciences*, *46*(8), 971–1010.

Lindsay, M. D., Jessell, M. W., Ailleres, L., Perrouty, S., de Kemp, E., & Betts, P. G. (2013). Geodiversity: Exploration of 3D geological model space. *Tectonophysics*, *594*, 27–37.

Lindsay, M. D., Perrouty, S., Jessell, M. W., & Ailleres, L. (2013). Making the link between geological and geophysical uncertainty: Geodiversity in the Ashanti Greenstone Belt. *Geophysical Journal International*, *195*(2), 903–922.

Lochbühler, T., Vrugt, J. A., Sadegh, M., & Linde, N. (2015). Summary statistics from training images as prior information in probabilistic inversion. *Geophysical Journal International*, *201*(1), 157–171.

Lomask, J., Guitton, A., Fomel, S., Claerbout, J., & Valenciano, A. A. (2006). Flattening without picking. *Geophysics*, *71*(4), 13–P20. https://doi.org/10.1190/1.2210848.

MacEachren, A. M. (1992). Visualizing uncertain information. *Cartographic Perspectives*, (13) 10–19-19.

MacEachren, M. A., Robinson, A., Hopper, S., Gardner, S., Murray, R., Gahegan, M., & Hetzler, E. (2005). Visualizing geospatial information uncertainty: What we know and what we need to know. *Cartography and Geographic Information Science*, *32*(3), 139–161.

MacKay, D. J. C. (2003). *Information theory, inference, and learning algorithms* (4th ed.). Cambridge, UK: Cambridge University Press.

Maerten, L., Gillespie, P., & Daniel, J.-M. (2006). Three-dimensional geomechanical modeling for constraint of subseismic fault simulation. *AAPG Bulletin*, *90*(9), 1337–1358. https://doi.org/10.1306/03130605148.

Mallet, J.-L. (1984). Automatic contouring in presence of discontinuities. In G. Verly, M. David, A. G. Journel, & A. Marechal (Eds.), *Geostatistics for natural resources characterization* (pp. 669–677). Dordrecht: Springer Netherlands.

Mallet, J. L. (1988). Three-dimensional graphic display of disconnected bodies. *Mathematical Geology*, *20*(8), 977–990. https://doi.org/10.1007/BF00892974.

Mallet, J.-L. (1989). Discrete smooth interpolation. *ACM Transactions on Graphics*, *8*(2), 121–144. https://doi.org/10.1145/62054.62057.

Mallet, J. L. (1992). GOCAD: A computer aided design program for geological applications. In A. K. Turner (Ed.), *Three-dimensional modeling with geoscientific information systems* (pp. 123–141). Dordrecht, Netherlands: Springer.
Mallet, J. L. (1997). Discrete modeling for natural objects. *Mathematical Geology, 29*(2), 199–219. https://doi.org/10.1007/BF02769628.
Mallet, J.-L. (2002). *Geomodeling*. Oxford University Press.
Mallet, J. L. (2004). Space–time mathematical framework for sedimentary geology. *Mathematical Geology, 36*(1), 1–32.
Mallet, J.-L. (2014). *Elements of mathematical sedimentary geology: The geochron model*. EAGE Publications.
Mallet, J.-L., & Tertois, A.-L. (2010). *Solid earth modeling and geometric uncertainties*. Society of Petroleum Engineers.
Mann, C. J. (1970). Randomness in nature. *GSA Bulletin, 81*(1), 95–104.
Mann, J. C. (1993). Uncertainty in geology. *Computers in geology—25 years of progress* (pp. 241–254). Oxford University Press, Inc.
Mäntylä, M. (1988). *An introduction to solid modeling*. Computer Science Press.
Marchal, D., Guiraud, M., Rives, T., & van den Driessche, J. (1998). Space and time propagation processes of normal faults. *Geological Society, London, Special Publications, 147*(1), 51–70. http://sp.lyellcollection.org/content/147/1/51.abstract.
Marechal, A. (1984). Kriging seismic data in presence of faults. In G. Verly, M. David, A. G. Journel, & A. Marechal (Eds.), *Geostatistics for natural resources characterization* (pp. 271–294). Dordrecht, Netherlands: Springer.
Mariethoz, G., & Caers, J. (2015). *Multiple-point geostatistics: Stochastic modeling with training images*. Chichester, West Sussex Hoboken, NJ: Wiley Blackwell.
Marjoram, P., Molitor, J., Plagnol, V., & Tavaré, S. (2003). Markov chain Monte Carlo without likelihoods. *Proceedings of the National Academy of Sciences, 100*(26), 15324–15328.
Martelet, G., Calcagno, P., Gumiaux, C., Truffert, C., Bitri, A., Gapais, D., & Brun, P. J. (2004). Integrated 3D geophysical and geological modelling of the hercynian suture zone in the champtoceaux area (South Brittany, France). *Tectonophysics, 382*(1–2), 117–128.
Martin, R., & Boisvert, J. B. (2017). Iterative refinement of implicit boundary models for improved geological feature reproduction. *Computers & Geosciences, 109*, 1–15. https://doi.org/10.1016/j.cageo.2017.07.003.
Matheron, G. (2012). *Estimating and Choosing: An Essay on Probability in Practice*. Berlin, Heidelberg: Springer.
Mavko, G., & Mukerji, T. (1998). A rock physics strategy for quantifying uncertainty in common hydrocarbon indicators. *Geophysics, 63*(6), 1997–2008.
Mavko, G., Mukerji, T., & Dvorkin, J. (2009). *The rock physics handbook. Tools for seismic analysis of porous media*. Cambridge: Cambridge University Press.
Maxelon, M., Renard, P., Courrioux, G., Brändli, M., & Mancktelow, N. (2009). A workflow to facilitate three-dimensional geometrical modelling of complex polydeformed geological units. *Computers & Geosciences, 35*(3), 644–658.
McCaffrey, K. J. W., Jones, R. R., Holdsworth, R. E., Wilson, R. W., Clegg, P., Imber, J., … Trinks, I. (2005). Unlocking the spatial dimension: Digital technologies and the future of geoscience fieldwork. *Journal of the Geological Society, 162*(6), 927–938.
McGaughey, J. (2006a). The common earth model: A revolution in mineral exploration data integration. *GIS for the Earth Sciences: Geological Association of Canada, Special Publication, 44*, 567–576.
McGaughey, J. (2006b). The Common Earth Model: A Revolution in Mineral Exploration Data Integration. In J. R. Harris (Ed.), *Special Publication. GIS for the Earth Sciences* (pp. 567–576). Geological Association of Canada.
Mello, U. T., & Henderson, M. E. (1997). Techniques for including large deformations associated with salt and fault motion in basin modeling. *Marine and Petroleum Geology, 14*(5), 551–564.

Merland, R., Caumon, G., Levy, B., & Collon-Drouaillet, P. (2014). Voronoi grids conforming to 3D structural features. *Computational Geosciences*, *18*(3–4), 373–383.
Ming, J., Pan, M., Qu, H., & Ge, Z. (2010). GSIS A 3D geological multi-body modeling system from netty cross-sections with topology. *Computers & Geosciences*, *36*(6), 756–767.
Moës, N., Gravouil, A., & Belytschko, T. (2002). Non-planar 3D crack growth by the extended finite element and level sets—Part I: Mechanical model: NON-planar 3D crack growth—Part I. *International Journal for Numerical Methods in Engineering*, *53*(11), 2549–2568. https://doi.org/10.1002/nme.429.
Montanari, A. (2007). What do we mean by 'uncertainty'? The need for a consistent wording about uncertainty assessment in hydrology. *Hydrological Processes*, *21*(6), 841–845.
Moorkamp, M., Heincke, B., Jegen, M., Roberts, A. W., & Hobbs, R. W. (2011). A framework for 3-D joint inversion of MT, gravity and seismic refraction data. *Geophysical Journal International*, *184*(1), 477–493.
Mosegaard, K., & Tarantola, A. (1995). Monte Carlo sampling of solutions to inverse problems. *Journal of Geophysical Research*, *100*(B7), 12–431.
Moyen, R., Mallet, J. L., Frank, T., Leflon, B., & Royer, J. J. (2004). 3D-parameterization of the 3D geological space—The geochron model. In *Proc. European conference on the mathematics of oil recovery (ECMOR IX)*.
Mukerji, T., Avseth, P., Mavko, G., Takahashi, I., & Gonzalez, E. F. (2001). Statistical rock physics: Combining rock physics, information theory, and geostatistics to reduce uncertainty in seismic reservoir characterization. *The Leading Edge*, *20*(3), 313–319.
Munthe, K., Holden, L., Mostad, P., & Townsend, C. (1994). Modelling sub-seismic Fault Patterns using a Marked Point Process. In *ECMOR IV—4th European conference on the mathematics of oil recovery*Norway: Røros. https://doi.org/10.3997/2214-4609.201411151.
Nearing, G. S., & Gupta, H. V. (2018). Ensembles vs. information theory: Supporting science under uncertainty. *Frontiers of Earth Science*, *90*(12), 1–8.
Nearing, G. S., Tian, Y., Gupta, H. V., Clark, M. P., Harrison, K. W., & Weijs, S. V. (2016). A philosophical basis for hydrological uncertainty. *Hydrological Sciences Journal*, *61*(9), 1666–1678.
Nilsson, B., Højberg, L. A., Refsgaard, C. J., & Troldborg, L. (2007). Uncertainty in geological and hydrogeological data. *Hydrology and Earth System Sciences*, *11*(5), 1551–1561.
Olea, R. A. (1974). Optimal contour mapping using universal kriging. *Journal of Geophysical Research*, *79*(5), 695–702.
Omre, H. (1987). Bayesian kriging—Merging observations and qualified guesses in kriging. *Mathematical Geology*, *19*(1), 25–39.
Osypov, K., Yang, Y., Fournier, A., Ivanova, N., Bachrach, R., Yarman, C. E., … Woodward, M. (2013). Model-uncertainty quantification in seismic tomography: Method and applications. *Geophysical Prospecting*, *61*(6), 1114–1134.
Pakyuz-Charrier, E., Lindsay, M., Ogarko, V., Giraud, J., & Jessell, M. W. (2018). Monte Carlo simulation for uncertainty estimation on structural data in implicit 3-D geological modeling, a guide for disturbance distribution selection and parameterization. *Solid Earth*, *9*(2), 385–402.
Pallero, J. L. G., Fernández-Martínez, J. L., Bonvalot, S., & Fudym, O. (2015). Gravity inversion and uncertainty assessment of basement relief via particle swarm optimization. *Journal of Applied Geophysics*, *116*, 180–191.
Pallero, J. L. G., Fernández-Martínez, J. L., Bonvalot, S., & Fudym, O. (2017). 3D gravity inversion and uncertainty assessment of basement relief via particle swarm optimization. *Journal of Applied Geophysics*, *139*, 338–350.
Pang, A. (2008). Visualizing uncertainty in natural hazards. In A. Bostrom, S. French, & S. Gottlieb (Eds.), *Risk assessment, modeling and decision support* (pp. 261–294). Berlin, Heidelberg: Springer.

Pang, A. T., Wittenbrink, C. M., & Lodha, S. K. (1997). Approaches to uncertainty visualization. *The Visual Computer*, *13*(8), 370–390.

Park, H., Scheidt, C., Fenwick, D., Boucher, A., & Caers, J. (2013). History matching and uncertainty quantification of facies models with multiple geological interpretations. *Computational Geosciences*, *17*(4), 609–621.

Parker, R. L. (1974). Best bounds on density and detph from gravity data. *Geophysics*, *39*(5), 644–649.

Pellerin, J., Caumon, G., Julio, C., Mejia-Herrera, P., & Botella, A. (2015). Elements for measuring the complexity of 3D structural models: Connectivity and geometry. *Computers & Geosciences*, *76*, 130–140.

Perrin, M., & Rainaud, J.-F. (2013). *Shared earth modeling: Knowledge driven solutions for building and managing subsurface 3D geological models*. Editions Technip.

Perrin, M., Zhu, B., Rainaud, J.-F., & Schneider, S. (2005). Knowledge-driven applications for geological modeling. *Journal of Petroleum Science and Engineering*, *47*(1–2), 89–104.

Polanyi, M. (1958). *Personal knowledge*. Routledge & Kegan Paul Ltd.

Polson, D., & Curtis, A. (2010). Dynamics of uncertainty in geological interpretation. *Journal of the Geological Society*, *167*(1), 5–10.

Potter, K., Gerber, S., & Anderson, E. W. (2013). Visualization of uncertainty without a mean. *IEEE Computer Graphics and Applications*, *33*(1), 75–79.

Potter, K., Kniss, J., Riesenfeld, R., & Johnson, C. R. (2010). Visualizing summary statistics and uncertainty. *Computer Graphics Forum*, *29*(3), 823–832.

Pouliot, J., Bédard, K., Kirkwood, D., & Lachance, B. (2008). Reasoning about geological space: Coupling 3D GeoModels and topological queries as an aid to spatial data selection. *Computers & Geosciences*, *34*(5), 529–541.

Pouzet, J. (1980). Estimation of a surface with known discontinuities for automatic contouring purposes. *Journal of the international association for mathematica geology*, *12*(6), 559–575. https://doi.org/10.1007/BF01034744.

Putz, M., Stuwe, K., Jessell, M. W., & Calcagno, P. (2006). Three-dimensional model and late stage warping of the plattengneis shear zone in the Eastern Alps. *Tectonophysics*, *412*(1–2), 87–103.

Pyrcz, M. J., & Deutsch, C. V. (2014b). *Geostatistical reservoir modeling*. Oxford University Press.

Pyrcz, M. J., & Deutsch, V. C. (2014a). *Geostatistical reservoir engineering*. Oxford: Oxford University Press.

Pyrcz, M. J., & White, C. D. (2015). Uncertainty in reservoir modeling. *Interpretation*, *3*(2), SQ7–SQ19.

Rankey, E. C., & Mitchell, J. C. (2003). That's why it's called interpretation Impact of horizon uncertainty on seismic attribute analysis. *The Leading Edge*, *22*(9), 820–828.

Refsgaard, C. J., van der Sluijs, P. J., Brown, J., & van der Keur, P. (2006). A framework for dealing with uncertainty due to model structure error. *Advances in Water Resources*, *29*(11), 1586–1597.

Renaudeau, J., Maerten, F., Malvesin, E., & Caumon, G. (2018). Implicit structural modeling with local meshless functions. *80th EAGE conference and exhibition 2018*. https://doi.org/10.3997/2214-4609.201801235.

Ringrose, P., & Bentley, M. (2015). *Reservoir model design*. Dordrecht, Netherlands: Springer.

Rivenæs, J. C., Otterlei, C., Zachariassen, E., Dart, C., & Sjøholm, J. (2005). A 3D stochastic model integrating depth, fault and property uncertainty for planning robust wells, Njord field, offshore Norway. *Petroleum Geoscience*, *11*(1), 57–65.

Røe, P., Abrahamsen, P., Georgsen, F., Syversveen, A. R., & Lia, O. (2010). *Flexible simulation of faults*. Society of Petroleum Engineers.

Røe, P., Georgsen, F., & Abrahamsen, P. (2014). An uncertainty model for fault shape and location. *Mathematical Geosciences*, *46*(8), 957–969.

Rongier, G., Collon-Drouaillet, P., & Filipponi, M. (2014). Simulation of 3D karst conduits with an object-distance based method integrating geological knowledge. *Geomorphology*, *217*, 152–164. https://doi.org/10.1016/j.geomorph.2014.04.024.
Rossignac, J. (2005). Shape complexity. *The Visual Computer*, *21*(12), 985–996.
Rowbotham, P., Kane, P., Bentley, M. (2010). Bias in geophysical interpretation—The case for multiple deterministic scenarios. *The Leading Edge*, *29*(5), 590–595.
Runge, A. K., Scherbaum, F., Curtis, A., & Riggelsen, C. (2013). An interactive tool for the elicitation of subjective probabilities in probabilistic seismic-hazard analysis. *Bulletin of the Seismological Society of America*, *103*(5), 2862–2874.
Salles, L., Ford, M., Joseph, P., Le Carlier De Veslud, C., & Le Solleuz, A. (2011). Migration of a synclinal depocentre from turbidite growth strata: The annot syncline, SE France. *Bulletin de la Societe Geologique de France*, *182*(3), 199–220. https://doi.org/10.2113/gssgfbull.182.3.199.
Saltus, R. W., & Blakely, R. J. (2011). Unique geologic insights from "non-unique" gravity and magnetic interpretation. *GSA Today*, *21*(12), 4–11.
Sambridge, M., Bodin, T., Gallagher, K., & Tkalcic, H. (2013). Transdimensional inference in the geosciences. *Philosophical Transactions of the Royal Society A: Mathematical, Physical and Engineering Sciences*, *371*(1984), 20110547. 20110547.
Samson, P., Dubrule, O., & Euler, N. (1996). *Quantifying the impact of structural uncertainties on gross-rock volume estimates*. Society of Petroleum Engineers.
Scheidt, C., & Caers, J. (2008). Representing spatial uncertainty using distances and kernels. *Mathematical Geosciences*, *41*(4), 397–419.
Schetselaar, E. M., Harris, J. R., Lynds, T., & de Kemp, E. A., (2007). Remote predictive mapping 1. Remote predictive mapping (RPM): A strategy for geological mapping of Canada's north. *Geoscience Canada*, *34*(3 and 4), 93–111.
Schmidt, S., Plonka, C., Götze, H. J., & Lahmeyer, B. (2011). Hybrid modelling of gravity, gravity gradients and magnetic fields. *Geophysical Prospecting*, *59*(6), 1046–1051.
Schneeberger, R., de la Varga, M., Egli, D., Berger, A., Kober, F., Wellmann, F., & Herwegh, M. (2017). Methods and uncertainty estimations of 3-D structural modelling in crystalline rocks: A case study. *Solid Earth*, *8*(5), 987–1002.
Schöniger, A., Wöhling, T., Samaniego, L., & Nowak, W. (2014). Model selection on solid ground: Rigorous comparison of nine ways to evaluate Bayesian model evidence. *Water Resour Res*, *50*(12), 9484–9513.
Schweizer, D., Blum, P., & Butscher, C. (2017). Uncertainty assessment in 3-D geological models of increasing complexity. *Solid Earth*, *8*(2), 515–530.
Seers, T. D., & Hodgetts, D. (2016). Probabilistic constraints on structural lineament best fit plane precision obtained through numerical analysis. *Journal of Structural Geology*, *82*, 37–47.
Seifert, D., Sonnenborg, T. O., Refsgaard, J. C., Højberg, A. L., & Troldborg, L. (2012). Assessment of hydrological model predictive ability given multiple conceptual geological models. *Water Resources Research*, *48*(6), W06503.
Seipel, S., & Lim, N. J. (2017). Color map design for visualization in flood risk assessment. *International Journal of Geographical Information Science*, *31*(11), 2286–2309.
Shannon, E. C. (1948). A mathematical theory of communication. *Bell System Technical Journal*, *27*, 379–423.
Shaw, J. H., Plesch, A., Tape, C., Suess, M. P., Jordan, T. H., Ely, G., … Munster, J. (2015). Unified structural representation of the southern california crust and upper mantle. *Earth and Planetary Science Letters*, *415*, 1–15.
Sibson, R. (1981). A brief description of natural neighbour interpolation. In V. Barnett (Ed.), *Interpolating multivariate data* (pp. 21–36). John Wiley & Sons.
Sides, E. J. (1997). Geological modelling of mineral deposits for prediction in mining. *Geologische Rundschau*, *86*(2), 342–353. https://doi.org/10.1007/s005310050145.
Silva, J. B. C., Medeiros, W. E., & Barbosa, V. C. F. (2001). Potential-field inversion: Choosing the appropriate technique to solve a geologic problem. *Geophysics*, *66*(2), 511–520.

Sivia, D. S. (2006). *Data analysis: A bayesian tutorial.* Oxford University Press.

Souche, L., Lepage, F., & Iskenova, G. (2013). Volume based modeling-automated construction of complex structural models. *75th EAGE conference & exhibition incorporating SPE EUROPEC 2013.*

Spiegelhalter, D., Pearson, M., & Short, I. (2011). Visualizing uncertainty about the future. *Science, 333*(6048), 1393–1400.

Sprague, K. B., & de Kemp, E. A. (2005). Interpretive tools for 3-D structural geological modelling Part II: Surface design from sparse spatial data. *GeoInformatica, 9*(1), 5–32. https://doi.org/10.1007/s10707-004-5620-8.

Srivastava, M. R. (1992). *Reservoir characterization with probability field simulation.*

Srivastava, R. M. (1994). *The interactive visualization of spatial uncertainty.* Society of Petroleum Engineers.

Srivastava, R. M., & Froidevaux, R. (2005). Probability field simulation: A retrospective. In *Geostatistics banff 2004* (pp. 55–64). Dordrecht: Springer.

Steno, N. (1669). *De solido intra solidium naturaliter contento dissertationis prodromus.* Florence.

Sukumar, S. R., Page, D. L., Koschan, A. F., & Abidi, M. A. (2008). Towards understanding what makes 3D objects appear simple or complex. In *2008 IEEE computer society conference on computer vision and pattern recognition workshops (CVPR Workshops)* (pp. 1–8). IEEE.

Suzuki, S., Caumon, G., & Caers, J. (2008). Dynamic data integration for structural modeling: Model screening approach using a distance-based model parameterization. *Computational Geosciences, 12*(1), 105–119.

Swanson, D. C. (1988). A new geological volume computer modeling system for reservoir description. In *SPE-17579-MS*SPE: Society of Petroleum Engineers. https://doi.org/10.2118/17579-MS.

Talwani, M. (1965). Computation with the help of a digital computer of magnetic anomalies caused by bodies of arbitrary shape. *Geophysics, 30*(5), 797–817.

Talwani, M., Worzel, J. L., & Landisman, M. (1959). Rapid gravity computations for two-dimensional bodies with application to the Mendocino submarine fracture zone. *Journal of Geophysical Research, 64*(1), 49–59.

Tarantola, A. (2005). *Inverse problem theory and methods for model parameter estimation.* Society for Industrial Mathematics.

Tarantola, A. (2006). Popper, Bayes and the inverse problem. *Nature Physics, 2*(8), 492.

Tarantola, A., & Valette, B. (1982). Generalized nonlinear inverse problems solved using the least squares criterion. *Reviews of Geophysics, 20*(2), 219–232.

Telford, W. M., Geldart, L. P., & Sheriff, R. E. (2009). *Applied geophysics.* Cambridge: Cambridge University Press.

Tertois, A. L., & Mallet, J. L. (2007). Editing faults within tetrahedral volume models in real time. *Geological Society, London, Special Publications, 292*(1), 89–101. https://doi.org/10.1144/SP292.5.

Thibert, B., Gratier, J. P., & Morvan, J. M. (2005). A direct method for modeling and unfolding developable surfaces and its application to the Ventura Basin (California). *Journal of structural geology, 27*(2), 303–316.

Thiele, S. T., Jessell, M. W., Lindsay, M., Ogarko, V., Wellmann, F., & Pakyuz-Charrier, E. (2016). The topology of geology 1: Topological analysis. *Jorunal of Structural Geology, 91*, 27–38.

Thiele, S. T., Jessell, M. W., Lindsay, M., Wellmann, F., & Pakyuz-Charrier, E. (2016). The topology of geology 2: Topological uncertainty. *Journal of Structural Geology, 91*, 74–87.

Thirion, J. P. (1998). Image matching as a diffusion process: an analogy with Maxwell's demons. *Medical Image Analysis, 2*(3), 243–260.

Thore, P. (2015). Uncertainty in seismic inversion: What really matters? *The Leading Edge, 34*(9), 1000–1004.

Thore, P., Shtuka, A., Lecour, M., Ait-Ettajer, T., & Cognot, R. (2002). Structural uncertainties: Determination, management, and applications. *Geophysics, 67*(3), 840–852.

Tikhonov, A. N. (1963). On the solution of ill-posed problems and the method of regularization. *Doklady Akademii Nauk SSSR, 151*(3), 501–504.
Troldborg, L., Refsgaard, C. J., Jensen, H. K., & Engesgaard, P. (2007). The importance of alternative conceptual models for simulation of concentrations in a multi-aquifer system. *Hydrogeology Journal, 15*(5), 843–860.
Tversky, A., & Kahneman, D. (1974). Judgment under uncertainty: Heuristics and biases. *Science, 185*(4157), 1124–1131.
Van der Wel, F. J. M., Van der Gaag, L. C., & Gorte, B. G. H. (1998). Visual exploration of uncertainty in remote-sensing classification. *Computers and Geosciences, 24*(4), 335–343.
Viard, T., Caumon, G., & Levy, B. (2011). Adjacent versus coincident representations of geospatial uncertainty: Which promote better decisions? *Computers & Geosciences, 37*(4), 511–520.
Visschers, V. H. M., Meertens, R. M., Passchier, W. W. F., & De Vries, N. N. K. (2009). Probability information in risk communication: A review of the research literature. *Risk Analysis, 29*(2), 267–287.
Vollgger, S. A., & Cruden, A. R. (2016). Mapping folds and fractures in basement and cover rocks using UAV photogrammetry, Cape Liptrap and Cape Paterson, Victoria, Australia. *Journal of Structural Geology, 85*, 168–187. https://doi.org/10.1016/j.jsg.2016.02.012.
Walters, R. F. (1969). Contouring by machine: A user's guide. *AAPG Bulletin, 53*, 2324–2340.
Wang, H., Wellmann, F., Li, Z., Wang, X., & Liang, R. Y. (2017). A segmentation approach for stochastic geological modeling using Hidden Markov random fields. *Mathematical Geosciences, 49*(2), 145–177.
Wang, X., & Hansen, R. O. (1990). Inversion for magnetic anomalies of arbitrary three-dimensional bodies. *Geophysics, 55*(10), 1321–1326. https://doi.org/10.1190/1.1442779.
Watson, D. (1999). The natural neighbor series manuals and source codes. *Computers & Geosciences, 25*(4), 463–466. https://doi.org/10.1016/S0098-3004(98)00150-2.
Wehr, H., Chevrot, S., Courrioux, G., & Guillen, A. (2018). A three-dimensional model of the Pyrenees and their foreland basins from geological and gravimetric data. *Tectonophysics, 734–735*, 16–32.
Weinzierl, W., Schulte, L., & Aarre, V. (2016). Volumetric impact of fault perturbation in the first Fresnel zone. *Interpretation, 4*(4), T419–T426. https://doi.org/10.1190/INT-2015-0192.1.
Wellmann, F. (2013). Information theory for correlation analysis and estimation of uncertainty reduction in maps and models. *Entropy, 15*(4), 1464–1485.
Wellmann, F., de la Varga, M., Murdie, R. E., Gessner, K., & Jessell, M. W. (2017). Uncertainty estimation for a geological model of the Sandstone greenstone belt, Western Australia—Insights from integrated geological and geophysical inversion in a Bayesian inference framework. *Geological Society, London, Special Publications, 453*, 41–52.
Wellmann, F., Horowitz, F. G., Schill, E., & Regenauer-Lieb, K. (2010a). Towards incorporating uncertainty of structural data in 3D geological inversion. *Tectonophysics, 490*(3–4), 141–151.
Wellmann, F., Horowitz, F. G., Schill, E., & Regenauer-Lieb, K. (2010b). Towards incorporating uncertainty of structural data in 3D geological inversion. *Tectonophysics, 490*(3–4), 141–151.
Wellmann, F., & Regenauer-Lieb, K. (2012). Uncertainties have a meaning: Information entropy as a quality measure for 3-D geological models. *Tectonophysics, 526*, 207–216.
Wellmann, J. F., Thiele, S. T., Lindsay, M. D., & Jessell, M. W. (2016). Pynoddy 1.0: An experimental platform for automated 3-D kinematic and potential field modelling. *Geoscientific Model Development, 9*(3), 1019–1035. https://doi.org/10.5194/gmd-9-1019-2016.
Wijns, C., Boschetti, F., & Moresi, L. (2003). Inverse modelling in geology by interactive evolutionary computation. *Journal of Structural Geology, 25*(10), 1615–1621.

Wijns, C., & Kowalczyk, P. (2007). Interactive geophysical inversion using qualitative geological constraints. *Exploration Geophysics*, *38*(3), 208.
Wildenschild, D., Vaz, C. M. P., Rivers, M. L., Rikard, D., & Christensen, B. S. B. (2002). Using X-ray computed tomography in hydrology: Systems, resolutions, and limitations. *Journal of Hydrology*, *267*(3–4), 285–297.
Won, I. J., & Bevis, M. (1987). Computing the gravitational and magnetic anomalies due to a polygon: Algorithms and Fortran subroutines. *Geophysics*, *52*(2), 232–238.
Wood, R., & Curtis, A. (2004). Geological prior information and its applications to geoscientific problems. *Geological Society, London, Special Publications*, *239*(1), 1–14.
Wu, X. (2017). Building 3D subsurface models conforming to seismic structural and stratigraphic features. *Geophysics*, *82*(3), IM21–IM30. https://doi.org/10.1190/geo2016-0255.1.
Wu, X., & Caumon, G. (2017). Simultaneous multiple well-seismic ties using flattened synthetic and real seismograms. *Geophysics*, *82*(1), IM13–IM20. https://doi.org/10.1190/geo2016-0295.1.
Wu, X., & Hale, D. (2015). Horizon volumes with interpreted constraints. *Geophysics*, *80*(2), IM21–IM33. http://library.seg.org/doi/abs/10.1190/geo2014-0212.1.
Yilmaz, O. (2001). *Seismic data analysis: Processing, inversion, and interpretation of seismic data.* SEG.
Zadeh, A. L. (1965). Fuzzy sets*. *Information and Control*, *8*(3), 338–353.
Zeng, H., Backus, M. M., Barrow, K. T., & Tyler, N. (1998). Stratal slicing; Part 1, Realistic 3-D seismic model. *Geophysics*, *63*(2), 502–513. https://doi.org/10.1190/1.1444351.
Zhang, Y., Zhong, D., Wu, B., Guan, T., Yue, P., & Wu, H. (2018). 3D parametric modeling of complex geological structures for geotechnical engineering of dam foundation based on T-splines: 3D parametric geological modeling based on T-splines. *Computer-Aided Civil and Infrastructure Engineering*, *33*(7), 545–570. https://doi.org/10.1111/mice.12343.
Zhao, J., Chen, S., Zuo, R., & Carranza, E. J. M. (2011). Mapping complexity of spatial distribution of faults using fractal and multifractal models: Vectoring towards exploration targets. *Computers and Geosciences*, *37*(12), 1958–1966.
Zheglova, P., Lelièvre, P. G., & Farquharson, C. G. (2018). Multiple level-set joint inversion of traveltime and gravity data with application to ore delineation: A synthetic study. *Geophysics*, *83*(1), R13–R30. https://doi.org/10.1190/geo2016-0675.1.
Zheglova, P., Lelièvre, P. G., & Farquharson, C. G. (2018). Multiple level-set joint inversion of traveltime and gravity data with application to ore delineation: A synthetic studyMultiple level-set joint inversion. *Geophysics*, *83*(1), R13–R30.
Zhu, H., Li, S., Fomel, S., Stadler, G., & Ghattas, O. (2016). A Bayesian approach to estimate uncertainty for full-waveform inversion using a priori information from depth migration. *Geophysics*, *81*(5), R307–R323.
Zhu, L., Zhang, C., Li, M., Pan, X., & Sun, J. (2012). Building 3D solid models of sedimentary stratigraphic systems from borehole data: An automatic method and case studies. *Engineering Geology*, *127*, 1–13.
Ziesch, J., Aruffo, C. M., Tanner, D. C., Beilecke, T., Dance, T., Henk, A., … Krawczyk, C. M. (2015). Geological structure and kinematics of normal faults in the Otway Basin, Australia, based on quantitative analysis of 3-D seismic reflection data. *Basin Research*, *29*, 129–148.
Zimmermann, H. J. (2000). An application-oriented view of modeling uncertainty. *European Journal of Operational Research*, *122*(2), 190–198.

CHAPTER TWO

Single-station polarization analysis applied to seismic wavefields: A tutorial

Stewart Greenhalgh[a], David Sollberger[b,c], Cedric Schmelzbach[b,*], Matthew Rutty[d]

[a]Department of Geosciences, KFUPM, Dhahran, Saudi Arabia
[b]Institute of Geophysics, ETH Zurich, Zurich, Switzerland
[c]Seismic Apparition, Zurich, Switzerland
[d]Petrosys, Adelaide, SA, Australia
*Corresponding author: e-mail address: cedric.schmelzbach@erdw.ethz.ch

Contents

Advances in Geophysics, Volume 59
ISSN 0065-2687
https://doi.org/10.1016/bs.agph.2018.09.002

Abstract

The polarization properties of seismic wavefields recorded by triaxial (three-component) stations can been exploited for event detection, seismic direction finding, and wavefield filtering. Analysis tools have been derived by extending conventional electromagnetic polarization theory to transient seismic wavelets. The basis of this class of polarization processing is a (complex) covariance matrix formed over a window of data to characterize single polarized events within random noise. An eigendecomposition of the (complex) covariance matrix provides eigenvalues and eigenvectors to describe the degree and properties of polarization as well as to formulate polarization filters. Different modes of polarized events require different analysis domains. Rectilinearly polarized events can be examined using a real-valued covariance matrix. In contrast, elliptically polarized events (Rayleigh waves) require the analysis of the analytic signal and a complex-valued covariance matrix to describe the elliptical motion. The success of the polarization measures depends on the signal-to-noise ratio. Furthermore, if multiple events interfere, the basic assumptions used for single-station polarization analysis are violated and the method breaks down. The choice of analysis window length controls the resolution and success of the analysis and depends on the dataset being examined. Ideally, the window should be as long as possible but must not include two (or more) separate events. Synthetic datasets, constructed for different modes of events and different noise levels, as well as real data from a seismic reflection and vertical seismic profiling (VSP) survey together with recordings of an earthquake are used to illustrate the benefits and limitations of polarization processing.

1. Introduction

Polarization analysis of single station, multicomponent seismic data have been performed with success by several researchers, predominantly in the field of earthquake seismology (Bataille & Chiu, 1991; Cichowicz, Green, & Brink, 1988; Flinn, 1965; Harris, 1990; Jarpe & Dowla, 1991; Jurkevics, 1988; Magotra, Ahmed, & Chael, 1987, 1989; Montalbetti & Kanasewich, 1970; Vidale, 1986). With the advent of three-component recording in seismic exploration, especially in vertical seismic profiling (VSP), polarization analysis has increasingly been used in the field of exploration seismology as well (Benhama, Cliet, & Dubeset, 1988; Esmersoy, 1984; Gaiser, 2016; Hearn & Hendrick, 1998; Knowlton & Spencer, 1996; Kulesh et al., 2007;

Rene et al., 1986; Shieh & Herrmann, 1990; Sriram et al., 1990; Winterstein & Meadows, 1991). The technique is mainly used to detect and identify specific wave types with different modes of vibration (e.g., P-, S-, and Rayleigh waves) as well as to devise filters to enhance or extinguish particular events (e.g., mute surface waves) and to provide a means of estimating the direction of particle motion for use in seismic direction finding and studies of S-wave birefringence of fractured reservoirs (De Meersman, Kendall, & Van der Baan, 2009). Polarization analysis is also useful when attempting to perform vector seismic imaging whereby the additional information obtained on direction of arrival can better constrain source location and reflector orientation. It also leads to higher resolution migration maps.

The first motion detected in any seismic record is usually assumed to indicate a P- or compressional wave having longitudinal rectilinear polarization. This fact underpins most attempts in VSP preprocessing to recover the orientation of a triaxial geophone in a cased borehole when gyro or other directional information is not available (Greenhalgh & Mason, 1995). Most earthquake triaxial location schemes also concentrate their direction finding efforts on the first few samples of the primary (direct) arrival. Some estimate bearing and elevation using a Monte Carlo search method, in which the solution space is systematically explored (Christoffserson, Husebye, & Ingate, 1988; Gal'perin, 1984; Jarpe & Dowla, 1991). Others use a covariance matrix eigendecomposition approach in which the principal eigenvector is used to estimate the polarization direction (Greenhalgh, Mason, & Zhou, 2005; Jackson, Mason, & Greenhalgh, 1991). Several other methods exist, but these are generally simplified versions of the two major methods mentioned earlier.

All single-station triaxial signal detection and direction finding techniques search for pure modes of vibration (Samson, 1983), assuming that any noise on the signal is random. If multiple events are present within the processing window, then this assumption is not valid and the analysis breaks down. This is why most direction finding algorithms are most successful on the cleaner fist arrivals. At later times of the seismic trace, there are often several superposed signals, producing a complex 3D particle motion.

In this chapter, we present and investigate the performance of the main numerical methods for polarization analysis based on the eigendecomposition of the (complex-valued) covariance matrix using synthetic and real data. The input synthetic three-component seismograms are generated as pure mode arrivals from a known direction, but contaminated by varying amounts of random noise. The effects of the analysis domain (time or frequency), window length, and signal-to-noise ratio (SNR) on the polarization processing technique can therefore be systematically investigated. This provides some guidance for the optimal choice of analysis parameters which may vary with the data under examination. The purpose is to examine the possibilities and limitations of polarization processing of seismic data using a single triaxial station.

2. The particle motion trajectory, hodograms, and the polarization ellipsoid

2.1 Three-dimensional particle motion

Born and Wolf (1975) give an excellent overview of the partial coherence and polarization of light and make use of the optical cross-power (coherency) matrix for determining ellipticity, eccentricity, and rotation angle. It is difficult to establish a complete analogy between optical and seismic concepts of polarization. Strictly speaking, polarization is a narrow-band phenomenon. Light is a continuous, harmonic wave whereas seismic waves are polychromatic (broadband) transients. The electric and magnetic fields are transverse to the propagation direction, with electric field polarization confined to a plane. Optical matrices are second rank; seismic are inherently third rank. The four most basic seismic wave types are (1) compressional or longitudinal (P-waves) with rectilinear polarization in the direction of propagation, (2) transverse or shear (S-waves) with rectilinear polarization perpendicular to the direction of propagation (in the plane of the wavefront), (3) Rayleigh surface waves having elliptical polarization in the plane formed by the vertical axis and the horizontal (radial) direction of propagation, and (4) Love surface waves having rectilinear polarization in the horizontal direction transverse to the propagation direction.

We begin our discussion with the specific case of particle motion along an elliptical trajectory. Fig. 1 shows schematically the polarization ellipse in the *X–Z* plane over one cycle of a harmonic seismic wave having polarization in the *X–Z* plane (e.g., a Rayleigh wave traveling in the *X* direction and recorded at the Earth's surface). The eccentricity of the ellipse as well as the tilt of the major axis are related to the amplitude and phase differences between the *X*- and *Z*-components. The resultant displacement vector both rotates and changes its amplitude as a function of time. It returns to its original position after one cycle. When the long axis is much greater than the short axis the eccentricity of the ellipse (equal to the ratio of semiminor to semimajor axis lengths) approaches zero and the motion is rectilinear. When the axis lengths are equal, then the eccentricity is 1 and we are dealing with circular polarization.

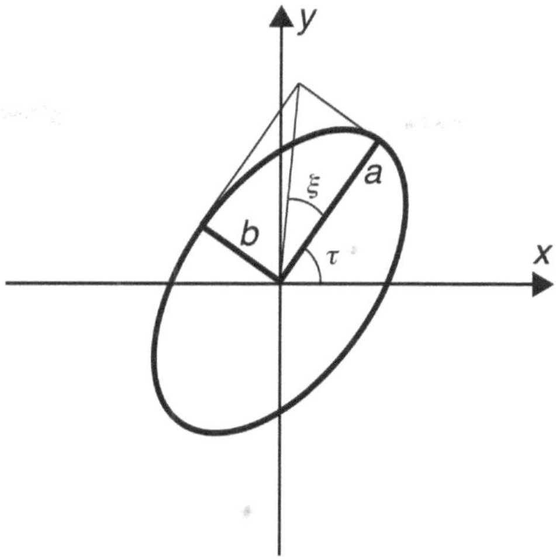

Fig. 1 The polarization ellipse showing how the displacement vector rotates and changes amplitude as a function of time over one period. The eccentricity (or ellipticity) e is the ratio of semimajor (a) to semiminor (b) axis length: $e = \frac{b}{a}$ and $\xi = \tan^{-1} e$. The ellipse is tilted away from the horizontal axis by an angle τ.

In general, the composite particle motion observed in a seismic record is a superposition of the various pure mode rectilinearly polarized arrivals (e.g., direct P-, direct S-, reflected P-, and reflected S-wave) as well as Rayleigh waves (elliptical polarization) and background noise (having isotropic polarization, which means random particle orientation or unpolarized). In general, the various waves (and noise) may overlap in time and the resultant motion (particle trajectory) is 3D and complex in shape (Cliet & Dubesset, 1987). Even pure modes can have their rectilinear polarization destroyed by phase change and mode conversion on reflection, associated with multi-pathing. Suppose we get two (or more) plane-polarized (elliptic) signals with different major and minor axes (alternatively, one elliptic signal plus noise) superimposed. The resulting motion as time elapses is 3D. The triaxial signal seems to stir around an ellipsoid in the field. In fact, the motion could as well trace out the shape of a four-bladed propeller, and it would, if the two arrivals remained constant. But in the case of transient signals they cannot. The propeller spins, and in the end what we see is the "equivalent ellipsoid," as the envelope of the spin.

2.2 Hodograms

Triaxial hodograms show the particle motion as a function of time in three orthogonal planes X–Y, X–Z, and Y–Z, over a specified time window. They show the phase relationships between traces but time is lost as a detailed diagnostic. Hodograms are useful for obtaining a good pictorial feel for the direction of polarization (or the direction of arrival in the case of P-waves) and the degree of rectilinearity of the particle motion. They indicate systematic changes in angle of arrival or sense of polarization, for example, a change from P- to S-wave.

Fig. 2 shows the hodograms in all three planes for a P-wave coming in at 70 degrees inclination and 20 degrees azimuth, an SV-wave arriving from the same direction, and a Rayleigh wave with a free parameter of $\pi/4$, coming in at 20 degrees azimuth (see Section 5 for a mathematical treatment of the polarization patterns). Rectilinear polarized events (P- and SV-wave),

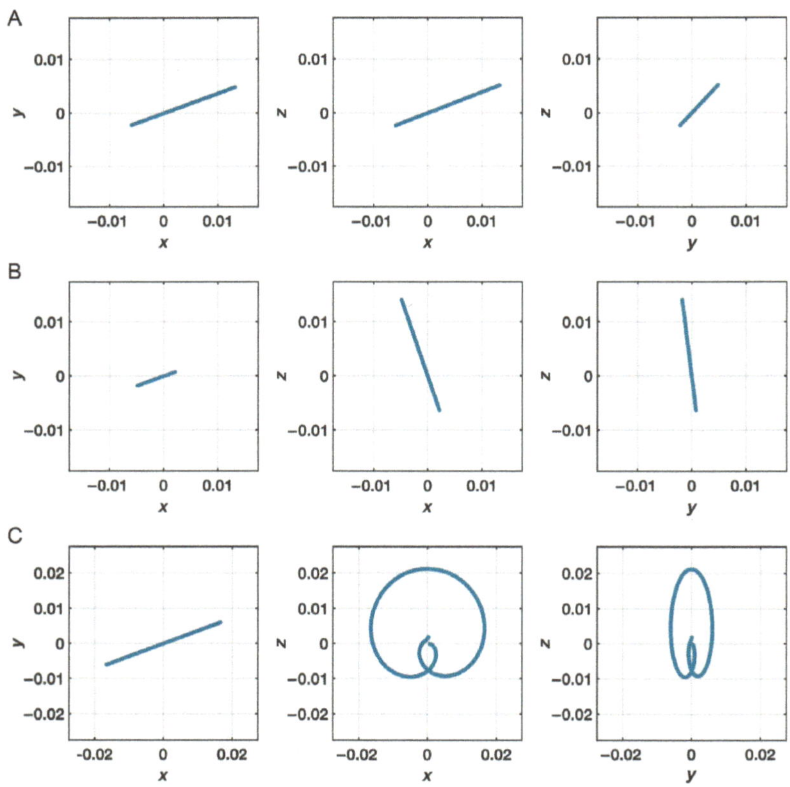

Fig. 2 Hodograms in all three planes (*X–Y*, *X–Z*, and *Y–Z*) for three separate arrivals: (A) P-wave (70 degrees inclination, 20 degrees azimuth), (B) SV-wave (70 degrees inclination, 20 degrees azimuth), and (C) Rayleigh wave (20 degrees azimuth, free parameter of the Rayleigh wave of $\pi/4$), showing the rectilinearity or ellipticity of the polarization. The direction of polarization can be gaged from the plots. The synthetic data were generated using a Ricker wavelet.

which display orthogonal polarization, can be easily distinguished from the elliptically polarized Rayleigh wave. The displays in each case are over one cycle. If the type of mode is known then it is also possible to make estimations of the direction angles directly from the hodogram display.

2.3 The polarization ellipsoid

We mentioned earlier that the seismic particle motion trajectory is in general quasi-ellipsoidal. The axes of the ellipsoid are directly related to the eigenvalues and eigenvectors of the 3×3 covariance matrix (see below, also Greenhalgh et al., 2005). The principal axis of the ellipsoid, corresponding to the dominant eigenvalue, defines the major polarization direction. For small phase differences between the three components of motion (X, Y, and Z), the ellipsoid is very elongate and it is possible to estimate with some certainty the direction of arrival if the dominant wave type is known. For zero phase difference between all components, such as with an isolated body wave in isotropic media, the motion is rectilinear. Most seismic direction finding methods assume quasilinear polarization.

Polarization is conveniently analyzed using the covariance matrix which is the outer product of the motion vector (X, Y, Z; e.g., corresponding displacement, particle velocity, or acceleration components) with itself. The matrix can be calculated for each discrete time in the window under analysis. The matrices each corresponding to a single time are then averaged over the time window to produce a single matrix containing *expectation* values. Thus

$$\mathbf{C} = \frac{1}{N}\sum_{i=1}^{N}\begin{pmatrix} X_i^2 & X_iY_i & X_iZ_i \\ Y_iX_i & Y_i^2 & Y_iZ_i \\ Z_iX_i & Z_iY_i & Z_i^2 \end{pmatrix}, \tag{1}$$

where i is the sample number and N is the number of samples in the time window. The sum needs to be over a sufficiently long time so that the average of each term approaches its expectation value (see Section 6). Eigenanalysis of the covariance matrix $\mathbf{C}$ (or its complex counterpart, sometimes referred to as the coherency matrix) provides a principal components analysis of the power. The power expectation in a triaxial trace can always be decomposed into three orthogonal (in time) components along three mutually perpendicular axes. The magnitude of these power components is given by the eigenvalues (λ_1, λ_2, and λ_3), and the unit vectors in the direction of the components are the corresponding eigenvectors ($\mathbf{e}_1$, $\mathbf{e}_2$, and $\mathbf{e}_3$). The eigenvectors define the axes of the polarization (covariance) ellipsoid (to be distinguished from the trajectory ellipsoid; see Cliet & Dubesset, 1987) and the eigenvalues define their relative lengths (see Jackson et al., 1991). A more detailed treatment will be given in Section 6 when working with the complex-valued covariance matrix.

2.4 Degree of polarization

Born and Wolf (1975, section 10.8.2) define a degree of polarization parameter for electromagnetic waves as the ratio of polarized intensity to total intensity:

$$P_B = \frac{I_{pol}}{I_{pol} + I_{noise}} = \frac{I_{pol}}{I_{total}}. \tag{2}$$

In the absence of a polarized signal, P_B tends to zero. If the signal is completely polarized P_B tends to unity.

A similar approach may be adopted for seismic waves using the eigendecomposition method.

If a dataset is purely polarized, then the (complex) covariance matrix will have only one nonzero eigenvalue. The degree of polarization may therefore be inferred by examining the eigenvalues of the (complex) covariance matrix.

Polarization has no meaning for a single time sample. It requires extended waveforms. The signal at a particular instant in time cannot give unambiguous information about the polarization of modes passing the triaxial station. Only through its association with adjacent samples can the polarization (and hence arrival directions) of modes be inferred. It is necessary to observe the behavior of the triaxial recording over a time window in order to obtain estimates for parameters describing the signal. On the other hand, it is necessary that these parameters remain (approximately) constant during that time, so they can be estimated and not some undetermined quantity. Polarization analysis requires that the relative phases and amplitudes of the three components are quasistatic. They must not change appreciably within the time window required to estimate the polarization. Multiple modes with different arrival times and arrival directions would produce changing relationships between the amplitudes and the phases of the three components. The concept of the polarization of a transient arrival only retains its meaning if one considers time windows short enough for the relationship between the amplitudes and the phases of the components to be approximately constant. We will examine the differences between monochromatic and transient signals later.

Estimates of the confidence of the identified events may then be made by forming picking functions related to the relative sizes of the three eigenvalues of the complex covariance matrix (see Section 6). Having sorted the eigenvalues, λ_i, into descending order ($\lambda_1 > \lambda_2 > \lambda_3$), several such functions are defined below:

$$P_1 = \frac{1}{2}\left[\frac{3\lambda_1}{\lambda_1 + \lambda_2 + \lambda_3} - 1\right] \tag{3}$$

(Esmersoy, 1984, eq. 13):

$$P_2 = \left[1 + \left(\frac{\varepsilon(\lambda_2 + \lambda_3)}{\lambda_1}\right)^{2N}\right]^{-1/2} \tag{4}$$

where N and ε are adjustable parameters, the Butterworth gain function (Esmersoy, 1984, eq. 18):

$$P_3 = 1 - \left(\frac{\lambda_2}{\lambda_1}\right)^N \tag{5}$$

(Montalbetti & Kanasewich, 1970, eq. 5):

$$P_4 = \frac{\lambda_1}{\lambda_2 + \lambda_3} \tag{6}$$

(Esmersoy, 1984, eq. 17). Another polarization parameter proposed by Esmersoy (1984) is

$$P_5 = \left(\frac{2\lambda_1}{\lambda_2 + \lambda_3} - 1\right)\sum_{i=1}^{3}\lambda_i \tag{7}$$

If there is no polarization, all three eigenvalues are equal to each other and P_5 becomes zero. Otherwise, P_5 increases linearly with increasingly linearly polarized power. Analogously to Eq. (2), the degree of polarization can be expressed by:

$$B = \frac{P_5}{P_5 + 1}. \tag{8}$$

where B is zero for unpolarized signals such as random noise and reaches unity for completely polarized signals. See Sections 8 and 9 for an application example of these polarization measures to synthetic and real data, respectively.

3. Polarization analysis of harmonic waves of arbitrary form

We begin by examining mathematically the classical monochromatic case before extending the analysis to a transient wavelet. The standard polarization analysis from optics (Born & Wolf, 1975) may be generalized to a situation more appropriate for seismic analysis. A rather general 3D harmonic wave of frequency ω has the form

$$\begin{aligned} V_x(\mathbf{r}, t) &= a_1(\mathbf{r})\cos[\omega t - \varphi_1(\mathbf{r})] \\ V_y(\mathbf{r}, t) &= a_2(\mathbf{r})\cos[\omega t - \varphi_2(\mathbf{r})] \\ V_z(\mathbf{r}, t) &= a_3(\mathbf{r})\cos[\omega t - \varphi_3(\mathbf{r})] \end{aligned} \tag{9}$$

where $\mathbf{r}$ is the position vector, t is the time, and V, a, and φ denote the particle displacement (or velocity or acceleration), amplitude, and phase, respectively, of the three components of motion. This may be written as:

$$\begin{pmatrix} V_x \\ V_y \\ V_z \end{pmatrix} = \begin{pmatrix} a_1\cos\varphi_1 \\ a_2\cos\varphi_2 \\ a_3\cos\varphi_3 \end{pmatrix}\cos(\omega t) + \begin{pmatrix} a_1\sin\varphi_1 \\ a_2\sin\varphi_2 \\ a_3\sin\varphi_3 \end{pmatrix}\sin(\omega t) \tag{10}$$

or

$$\begin{aligned} \mathbf{V}(\mathbf{r}, t) &= \mathbf{p}(\mathbf{r})\cos(\omega t) + \mathbf{q}(\mathbf{r})\sin(\omega t) \\ &= \mathrm{Re}\{\mathbf{U}(\mathbf{r})\mathrm{e}^{-i\omega t}\} \end{aligned} \tag{11}$$

where $\mathrm{Re}\{\mathbf{U}\}$ is the real part of the complex vector $\mathbf{U}$

$$\mathbf{U}(\mathbf{r}) = \mathbf{p}(\mathbf{r}) + i\mathbf{q}(\mathbf{r}). \tag{12}$$

The motion described by Eqs. (10) and (11) as time varies is an ellipse (2D motion). This is illustrated by considering orthogonal vectors **f** and **g**, chosen to satisfy

$$\mathbf{p} + i\mathbf{q} = (\mathbf{f} + i\mathbf{g})\mathrm{e}^{i\xi}. \tag{13}$$

On substitution in Eq. (11) **V** becomes

$$\begin{aligned} \mathbf{V} &= \Re\left\{[\mathbf{f} + i\mathbf{g}]\mathrm{e}^{-i(\omega t - \xi)}\right\} \\ &= \mathbf{f}\cos(\omega t - \xi) + \mathbf{g}\sin(\omega t - \xi) \end{aligned} \tag{14}$$

Choosing Cartesian axes along **f**, **g**, and $(\mathbf{f}\times\mathbf{g})$ rotate the coordinate frame of reference into the principal axes of the ellipse with new components:

$$\mathbf{V}_1=|\mathbf{f}|\cos(\omega t-\xi),\quad \mathbf{V}_2=|\mathbf{g}|\sin(\omega t-\xi),\quad \mathbf{V}_3=0. \tag{15}$$

This gives the equation of the polarization ellipse

$$\frac{V_1^2}{f^2}+\frac{V_2^2}{g^2}=1 \tag{16}$$

with its semimajor and semiminor axes lengths, f and g, given by:

$$\begin{aligned}|f|^2&=\frac{1}{2}\left[p^2+q^2+\sqrt{(p^2-q^2)^2+4(\mathbf{p}.\mathbf{q})^2}\right]\\ |g^2|&=\frac{1}{2}\left[p^2+q^2-\sqrt{(p^2-q^2)^2+4(\mathbf{p}.\mathbf{q})^2}\right]\end{aligned} \tag{17}$$

A sufficient condition for linear polarization (the degenerate case) is $g=0$. This condition is determined by the relative phase terms of each component of motion and implies:

$$a_1^2a_2^2\sin^2(\varphi_1-\varphi_2)+a_2^2a_3^2\sin^2(\varphi_2-\varphi_3)+a_3^2a_1^2\sin^2(\varphi_3-\varphi_1)=0. \tag{18}$$

As in the 2D analysis of Born and Wolf (1975), the special case of linearly polarized monochromatic motion implies that each component must be either in-phase or have a phase shift which is an integer multiple of π:

$$\begin{aligned}\varphi_2-\varphi_1&=k\pi\\ \varphi_3-\varphi_2&=l\pi\\ \varphi_1-\varphi_3&=m\pi\end{aligned} \tag{19}$$

with k, l, and m being integers and $k+l+m=0$ (since the left-hand side terms sum to zero). The linear direction of particle motion is now defined by the vector

$$\mathbf{D}=\begin{pmatrix}(-1)^k a_1\\ (-1)^l a_2\\ (-1)^m a_3\end{pmatrix} \tag{20}$$

in the coordinate axis frame of recording.

In summary, we note that any rectilinear monochromatic motion recorded by a triaxial (three-component; 3C) detector should display specific phase shifts (integer multiples of π) between components. Polarization analysis therefore uses the relative phase differences of the components of motion to detect rectilinear motion and the relative amplitudes of the motion on each component to then infer the principal direction of particle vibration.

In the above analysis, we assumed that the amplitudes a and phases φ were not changing with time but were quasistatic. If they are time varying then the particle motion becomes 3D rather than 2D and the concept of polarization partly loses its meaning.

4. Transient wavelets

An arbitrary waveform may be expressed as the Fourier sum of such harmonic waves as described earlier. It is useful to consider quasi-monochromatic waves as a first approximation to a transient wavelet. A wave is said to be quasi-monochromatic when its spectral bandwidth, Δf, is small compared to the time of its observation, T. This leads to the concept of a characteristic coherence time $\Delta t = \frac{1}{\Delta f}$ when measuring the wavefield as it passes a fixed position in space. There is, therefore, a time window of coherence around each time sample of the trace, dependent on the spectral bandwidth of the signal. Within this window, the complex amplitude and phase of the signal, and hence its polarization properties, remain relatively constant and may therefore be estimated by averaging over the complete window (an integral measure). Using a time window over the data which is shorter than the coherence time is the basis of the covariance matrix techniques referred to above and described in detail in a later section. As the spectral bandwidth of the signal increases, so the time window for the analysis should shorten.

Another approach for examining a transient seismic event is to decompose it into narrow frequency bands, which bracket the spectrum of interest. The data vector may then be analyzed within each frequency band to see if any significant polarization characteristics are present. A transient wavelet will be linearly polarized if all its Fourier components are linearly polarized in the same direction. Note that in this case the complex amplitude and the phase of the signal do not necessarily have to remain slowly varying with time (as in the quasi-monochromatic case) but the relative amplitude and the phase differences between each component do. Linear polarization is observable in the time domain. Each component in space records the same transient waveform multiplied by a scalar value proportional to the direction ratios of the polarization direction. The use of a measure to detect constant amplitude and phase differences between components within a time window is therefore the way polarization processing is performed. Such measures are the real and complex covariance matrices described below.

5. Pure mode arrivals and the polarization model

5.1 Body waves within a medium

The expected particle polarization for perfect single mode elastic arrivals is either 1D (e.g., rectilinear P- or S-waves) or 2D (e.g., plane-polarized Rayleigh waves). An event is purely polarized if

$$\mathbf{u}(\omega) = \mathbf{h}_i s(\omega) \tag{21}$$

for all frequencies, where $\mathbf{u}$ is the n-dimensional signal recorded at the multicomponent station, s is a scalar wavelet function describing the transient event, and $\mathbf{h}_i$ is the constant (complex) polarization vector. The $\mathbf{h}_i$ is also called the pure states of vibration (Samson, 1983).

In a 3D space, any pure mode arrival direction may be described in terms of its polar angles, θ and ϕ, denoting inclination to the vertical and the azimuth, respectively (Fig. 3). Esmersoy (1984) defines the polarization

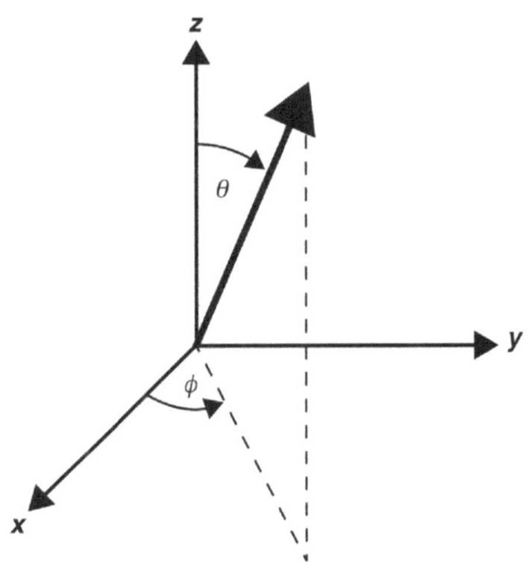

Fig. 3 Right-handed coordinate frame considered in this chapter.

vectors for the rectilinearly polarized P-, SV-, and SH-modes. These are given below for the (x, y, z) coordinate frame (z-positive upward, see Fig. 3):

$$\mathbf{h}_{\mathrm{P}} = \begin{bmatrix} \sin\theta\cos\phi \\ \sin\theta\sin\phi \\ \cos\theta \end{bmatrix}, \quad \mathbf{h}_{\mathrm{SH}} = \begin{bmatrix} -\sin\phi \\ \cos\phi \\ 0 \end{bmatrix}, \quad \mathbf{h}_{\mathrm{SV}} = \begin{bmatrix} -\cos\theta\cos\phi \\ -\cos\theta\sin\phi \\ \sin\theta \end{bmatrix} \tag{22}$$

5.2 Body waves at the free surface and surface waves

The displacements measured at the free surface due to an incoming P-wave (P in Fig. 4A) consists of the superposition of the upgoing P-wave, impinging at inclination θ_{P}, the reflected downgoing P-wave (marked as PP in Fig. 4A), and the mode-converted downgoing SV-wave (marked as SV in Fig. 4A with inclination θ_{S}). Similarly, the displacement due to an incoming SV-wave at the free surface (Fig. 4B) is the superposition of the upgoing SV-wave, impinging at inclination θ_{S}, the reflected downgoing SV-wave, and the mode-converted downgoing *P*-wave with inclination θ_{P}. As a consequence of the free surface effect, the apparent polarization angles $\theta_{\mathrm{Ap}}\,\theta_{\mathrm{As}}$ observed at the seismic station do not correspond to the true incident angles θ_{P} and θ_{S} of the P- and S-waves. For example, the apparent P-wave

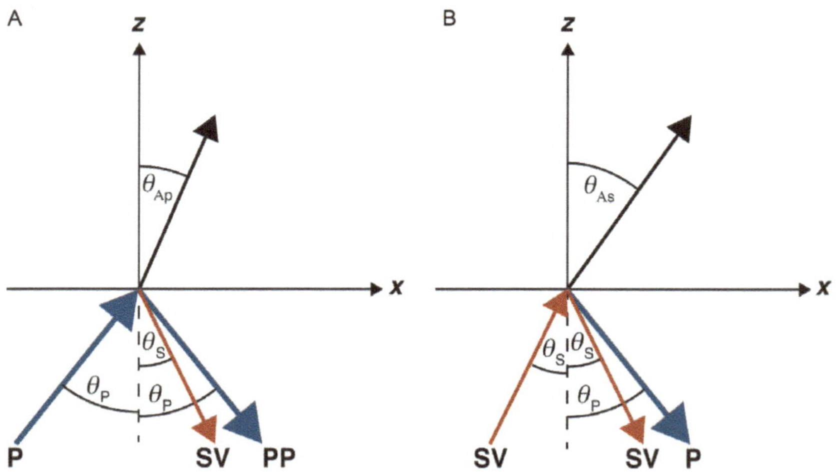

Fig. 4 (A) Free surface effect for an incident P-wave. The wavefield observed at the free surface (*black arrow*) is the superposition of the incident P-wave (P; *blue arrow*), the downgoing reflected P-wave (PP; *blue arrow*), and the mode-converted S-wave (SV; *red arrow*). (B) The free surface effect for an incident SV-wave. In (A) and (B), θ_{Ap} and θ_{Ap} mark the apparent P- and S-wave incident angles, respectively, θ_{P} is the P-wave incidence angle, and θ_{S} is the S-wave incidence angle.

incidence angle corresponds to twice the angle of the departing (reflected) SV-wave (e.g., Greenhalgh, Mason, Lucas, Pant, & Eames, 1990):

$$\theta_{\mathrm{As}} = 2\theta_{\mathrm{S}} \tag{23}$$

For incident SV-waves, the composite motion is more complicated (see Nuttli, 1961) and becomes elliptical at incident angles θ_{S} greater than the critical angle $\sin^{-1}(V_{\mathrm{S}}/V_{\mathrm{P}})$, where V_{S} and V_{P} are the shear and compressional wave velocities, respectively, of the near-surface material.

Note that SH-waves are decoupled from P-waves and hence no mode conversions occur when an SH-wave impinges at the free surface. An SH-wave recording at the free surface consists therefore only of the superposition of the upgoing SH-wave (impinging at inclination θ_{S}) and the reflected downgoing SH-wave.

The polarization vectors for the P-, SV-, and SH-modes at the free surface are given below (see Appendix for the amplitudes of the incident and reflected waves A_{P}, A_{PP}, A_{PS}, A_{S}, A_{SS}, and A_{SP}; Achenbach, 1973):

$$\mathbf{h}_{\mathrm{p}}^{\mathrm{FS}} = \begin{bmatrix} \sin\theta_{\mathrm{P}}\cos\phi + \dfrac{A_{\mathrm{PP}}}{A_{\mathrm{P}}}\sin\theta_{\mathrm{P}}\cos\phi + \dfrac{A_{\mathrm{PS}}}{A_{\mathrm{P}}}\cos\theta_{\mathrm{S}}\cos\phi \\ \sin\theta_{\mathrm{P}}\sin\phi + \dfrac{A_{\mathrm{PP}}}{A_{\mathrm{P}}}\sin\theta_{\mathrm{P}}\sin\phi + \dfrac{A_{\mathrm{PS}}}{A_{\mathrm{P}}}\cos\theta_{\mathrm{S}}\sin\phi \\ \cos\theta_{\mathrm{P}} - \dfrac{A_{\mathrm{PP}}}{A_{\mathrm{P}}}\cos\theta_{\mathrm{P}} + \dfrac{A_{\mathrm{PS}}}{A_{\mathrm{P}}}\sin\theta_{\mathrm{S}} \end{bmatrix},$$

$$\mathbf{h}_{\mathrm{SH}}^{\mathrm{FS}} = \begin{bmatrix} -2\sin\phi \\ 2\cos\phi \\ 0 \end{bmatrix}, \tag{24}$$

$$\mathbf{h}_{\mathrm{SV}}^{\mathrm{FS}} = \begin{bmatrix} -\cos\theta_{\mathrm{S}}\cos\phi + \dfrac{A_{\mathrm{SS}}}{A_{\mathrm{S}}}\cos\theta_{\mathrm{S}}\cos\phi + \dfrac{A_{\mathrm{SP}}}{A_{\mathrm{S}}}\sin\theta_{\mathrm{P}}\cos\phi \\ -\cos\theta_{\mathrm{S}}\sin\phi + \dfrac{A_{\mathrm{SS}}}{A_{\mathrm{S}}}\cos\theta_{\mathrm{S}}\sin\phi + \dfrac{A_{\mathrm{SP}}}{A_{\mathrm{S}}}\sin\theta_{\mathrm{P}}\sin\phi \\ \sin\theta_{\mathrm{S}} + \dfrac{A_{\mathrm{SS}}}{A_{\mathrm{S}}}\sin\theta_{\mathrm{S}} - \dfrac{A_{\mathrm{SP}}}{A_{\mathrm{S}}}\cos\theta_{\mathrm{P}} \end{bmatrix},$$

with the superscript FS marking the free surface polarization vectors. The plane-polarized Rayleigh and Love modes have polarization vectors given by:

$$\mathbf{h}_{\mathrm{R}} = \begin{bmatrix} j\sin\xi\cos\phi \\ j\sin\xi\sin\phi \\ \cos\xi \end{bmatrix}, \quad \mathbf{h}_{\mathrm{L}} = \begin{bmatrix} \sin\phi \\ -\cos\phi \\ 0 \end{bmatrix}. \tag{25}$$

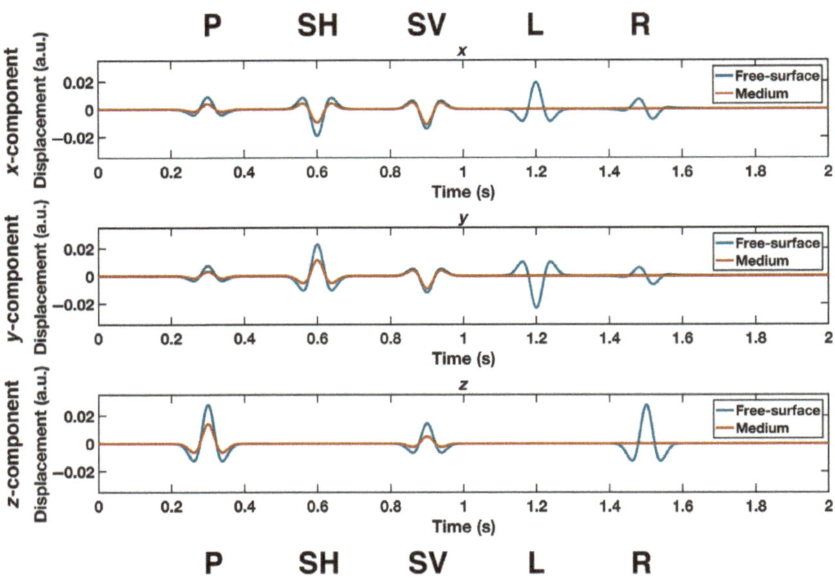

Fig. 5 Pure modes of polarization for a seismic signal (P-, SV-, SH-, Love, and Rayleigh waves) as observed on a triaxial station within the far interior of a medium (*red lines*) and at the free surface (*blue lines*).

In Eq. 25, $\xi = \tan^{-1} e$, where e is ellipticity, ξ is the free parameter of the Rayleigh mode, and j is the imaginary unit with the characteristic identity $j^2 = -1$. Eq. (25) implies that the phase shift between vertical and horizontal components is $\pi/2$ and therefore the polarization ellipse ought to have its major axis vertical for such an idealized representation. The pure modes observed within a medium and at the free surface are illustrated in graphical form in Fig. 5.

6. Polarization processing using the complex covariance matrix

6.1 The complex covariance or coherency matrix

The use of the complex covariance (or coherency) matrix for polarization analysis is documented by Vidale (1986). The digitized data vector $\mathbf{u}(t) = x(t)\hat{\mathbf{i}} + y(t)\hat{\mathbf{j}} + z(t)\hat{\mathbf{k}}$, where $\hat{\mathbf{i}}, \hat{\mathbf{j}},$ and $\hat{\mathbf{k}}$ are unit vectors in the Cartesian coordinate frame, is first converted to its analytic equivalent using the Hilbert transform

$$\mathbf{d} = \mathbf{u}(t) + j\mathrm{H}[\mathbf{u}(t)], \tag{26}$$

where H[ ] is the Hilbert transform and **d** is the analytic data vector. The Hilbert transformed signal corresponds to a $\pi/2$-phase shifted version of the original signal. The complex covariance or coherency matrix, **C**, for a particular time window of the analytic data, centered at t_0, is formed by taking the sum of the outer products of the analytic signal with its complex conjugate within the window:

$$\mathbf{C}[t_0] = \sum_{t_0-\frac{T}{2}}^{t_0+\frac{T}{2}} \mathbf{d}(t)\mathbf{d}^*(t) \tag{27}$$

where T is the time window length and * represents complex conjugate transpose operator. This matrix is a square $n \times n$ matrix, where n is the dimension of the recording geometry. When observing a seismic wavefield at a fixed position in space with a single three-component geophone station, n is equal to 3. Each element (α, β) within the complex covariance matrix is a zero-lag cross-correlation term for the two channels, α and β.

For successful polarization processing, the time window for the analysis should ideally contain only a single event. The directional information of a polarized arrival within the window then constructively adds in the complex covariance (or coherency) matrix while any random noise tends to cancel. Therefore, any single polarized event will be enhanced within the complex covariance matrix and should be "detected" by a drop in rank (or dimensionality) of the complex covariance matrix. An eigenanalysis of the matrix indicates any linear dependence between channels of the data. If such a dependence exists (due to the presence of a polarized signal), it will therefore be indicated by a dominant eigenvalue. The corresponding eigenvector provides the polarization direction, making the eigenanalysis of the complex covariance matrix a very useful tool for polarization processing.

6.2 Effect of random noise and interfering events

Seismic data are most unlikely to be purely polarized. Causes include the presence of random noise, the limited bandwidth of the signal, and the fact that the wavefield is not stationary. An analytic approach is used below to develop guidelines for applying polarization processing to such wavefields.

The recorded data, $\mathbf{u}(t)$, at a single station is a three-component vector. Consider the case of a transient rectilinear arrival. The data recorded can then be represented as

$$\mathbf{d} = \mathbf{S}_1(t)\hat{\mathbf{p}} + \mathbf{n}(t) \tag{28}$$

where $\mathbf{d}$ is the analytic data from $\mathbf{u}$, $\mathbf{S}_1(t)$ defines its analytic wavelet, $\mathbf{n}$ is the analytic noise vector and

$$\hat{\mathbf{p}} = \{p_x \; p_y \; p_z\}^T \tag{29}$$

are the direction cosines of the rectilinear polarization event. Taking a window length T centered on an arrival at time t_0, the complex covariance (coherency) matrix formed from this data is

$$\begin{aligned}\mathbf{C}[t_0] = \frac{1}{T}\int_{t_0-\frac{T}{2}}^{t_0+\frac{T}{2}} \mathbf{S}_1(t)\mathbf{S}_1^*(t)\hat{\mathbf{p}}\hat{\mathbf{p}}^*\,\mathrm{d}t + \frac{1}{T}\int_{t_0-\frac{T}{2}}^{t_0+\frac{T}{2}} \mathbf{S}_1(t)\hat{\mathbf{p}}\mathbf{n}^*(t)\,\mathrm{d}t \\ + \frac{1}{T}\int_{t_0-\frac{T}{2}}^{t_0+\frac{T}{2}} \mathbf{S}_1^*(t)\mathbf{n}(t)\hat{\mathbf{p}}^*\,\mathrm{d}t + \frac{1}{T}\int_{t_0-\frac{T}{2}}^{t_0+\frac{T}{2}} \mathbf{n}(t)\mathbf{n}^*(t)\,\mathrm{d}t\end{aligned} \tag{30}$$

where * represents the conjugate transpose operator.

Consider the structure of the complex covariance matrix. The polarized motion within the time window will constructively add, improving the ratio of signal to random noise. If the ratio of signal and random noise energy is sufficient for the window chosen, the direction of the polarization state vector of the signal will dominate within the complex covariance matrix. A choice of $N=1$ (a single sample) is clearly inadequate because each vector is a fully polarized signal in terms of this analysis even if there is no signal at all. The signal and noise are indistinguishable for small window lengths. The impact of random noise is discussed further in the next section.

6.2.1 Increasing SNR

If the SNR is large, the first term on the right-hand side of Eq. (30) dominates. This term is a scalar, determined by the analytic signal within the time window, multiplied by the matrix generated as the outer product of the motion vector with itself. The complex covariance matrix is therefore approximately rank 1 and its eigendecomposition will reveal the signal direction, $\mathbf{p}$, in the eigenvector corresponding to the dominant eigenvalue. The eigenvalue will be approximately equal to the above scalar and, therefore, includes phase information related to any frequency differences between the different components and the time over which the signal is averaged. It is therefore clear that the amplitudes of the complex covariance matrix elements are dependent on the positioning and length of the time window, as one would expect for a transient event. *Successful polarization analysis is therefore dependent on the choice of the time window.*

Consider now the effects of increasing noise in Eq. (30). As the SNR decreases toward unity, the second to fourth term in Eq. (30) become significant. However, if the noise is purely random, then the second and third terms tend to zero as the window length increases. This indicates one of the major strengths of single-station polarization analysis: *purely random noise cancels itself over the time window used for the real and complex covariance matrix.*

The eigenvector corresponding to the principal eigenvalue is then the polarization direction of the coherent signal. Most single-station polarization processing therefore assumes that any noise within the time window of analysis is random. If this criterion is satisfied, then the technique is highly successful. Indeed, if the noise is purely random, the only directional information in the complex covariance matrix is due to the signal (the first term on the RHS of Eq. 28), irrespective of the noise, as shown by Greenhalgh, Zhou, and Rutty (2008). However, in practice, noise is most unlikely to be completely random, so the SNR level in the processed time window is important.

Harris (1990) has compared the performance of a three-component station in varying levels of noise. He concluded that slowness (inclination angle) errors appear to be dominated by noise below a power SNR of around 10 for his data. A useful observation in his results is the way increasing noise levels degrade the predicted polarization direction gradually. He used the Cramer-Rao bound (see Esmersoy, 1984) to obtain lower bounds on the three-component single-station directional performance down to SNR values of about 1.0. This bound indicates that root mean square (rms) error reduces inversely as the SNR increases.

6.2.2 Effect of window length

For signals above a certain threshold, we would expect the polarization direction error to decrease as both the window length (time–bandwidth product) and the SNR increase:

$$\eta \approx \frac{1}{\sqrt{N}} \frac{1}{\text{SNR}} \tag{31}$$

Raising N helps the off-diagonal noise–signal products in the covariance matrix to random walk toward zero. Noise powers add, signals reinforce linearly for as long as the transient signal exists. The window length should be chosen such that it is as long as possible, without causing separate events on the record to interfere. The optimum choice of window length is therefore subjective. There is obviously a trade-off between stability of direction estimates and resolution of different events. We recommend that the window length be chosen such that it is as long as possible without causing separate events to interfere. The minimum window length should be at least half the dominant period of any expected events to ensure reasonable stability of directional estimates. This obviously means that events separated by less than a time equal to the sum of this period and their duration will interfere. If such interference is present, then single-station polarization processing is inappropriate.

From numerical tests (not shown), we found the errors to be substantial for window lengths less than half the pulse period, but good to within a few degrees for SNR >3 for windows longer than the pulse period.

6.2.3 Multiple events and coherent noise

A fundamental problem exists if there are overlapping events in the analysis window. Multiple events interfere with each other making polarization processing inappropriate on a single station. The problem may be visualized using a simple hodogram analysis. Three time windows of the synthetic dataset illustrated in Fig. 6 have been chosen for hodogram analysis. The first window (Fig. 6B) contains only random noise (isotropically polarized) and, hence, no apparent dominant direction. Each of the 2D plots is scattered. The second window (Fig. 6C) shows a strongly polarized rectilinear event (i.e., a P-wave), clearly indicated by the linearity of the hodogram, the slope of which provides the polarization direction. Note that the three eigenvalues of the complex covariance matrix denoted by λ_i, in descending order (see Section 2.3), are such that $\lambda_1 \gg \lambda_2, \lambda_3$ (Fig. 7). The third window (Fig. 6D) contains two interfering linearly polarized events (i.e., a P-wave and an SH-wave). The resulting motion is no longer polarized and cannot be detected using conventional polarization analysis.

Single-station polarization processing seeks for a single event in the analysis window. When two rectilinearly polarized events overlap they interfere, producing composite particle motion (see Fig. 6D). In general, none of the principal directions obtained from an eigenanalysis will then point along the direction vectors of the individual event directions, which is also reflected by the two significant eigenvalues λ_1, λ_2 (Fig. 7). All that is then known is that the original direction vectors of the two signals both lie in the plane spanned by the first two principal directions (eigenvectors corresponding to the two largest eigenvalues) of the covariance matrix.

Coherent noise is a fundamental problem in a standard single-station polarization analysis and the use of a model defined by Eq. (28) places

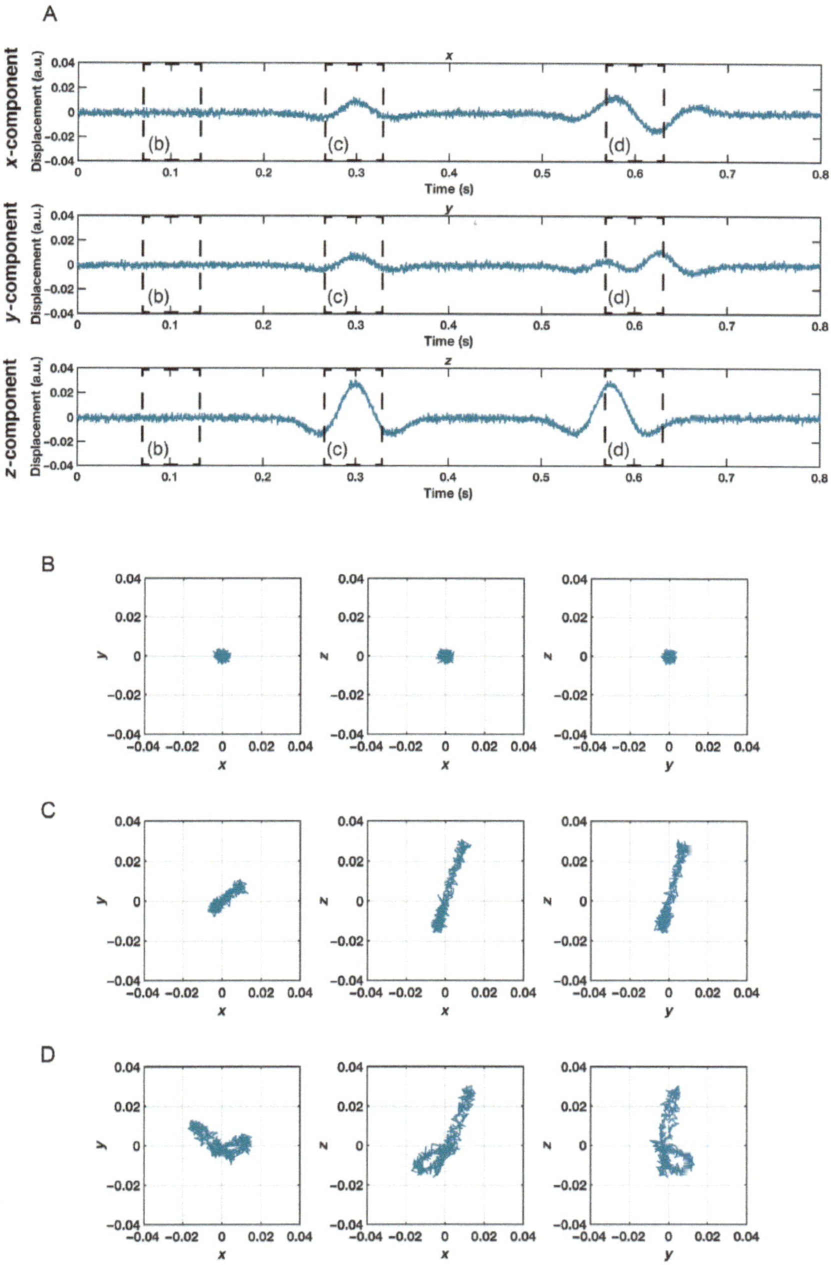

Fig. 6 (A) A three-component trace set illustrating hodograms for (B) random noise; (C) a single rectilinearly polarized event; and (D) two overlapping rectilinear events.

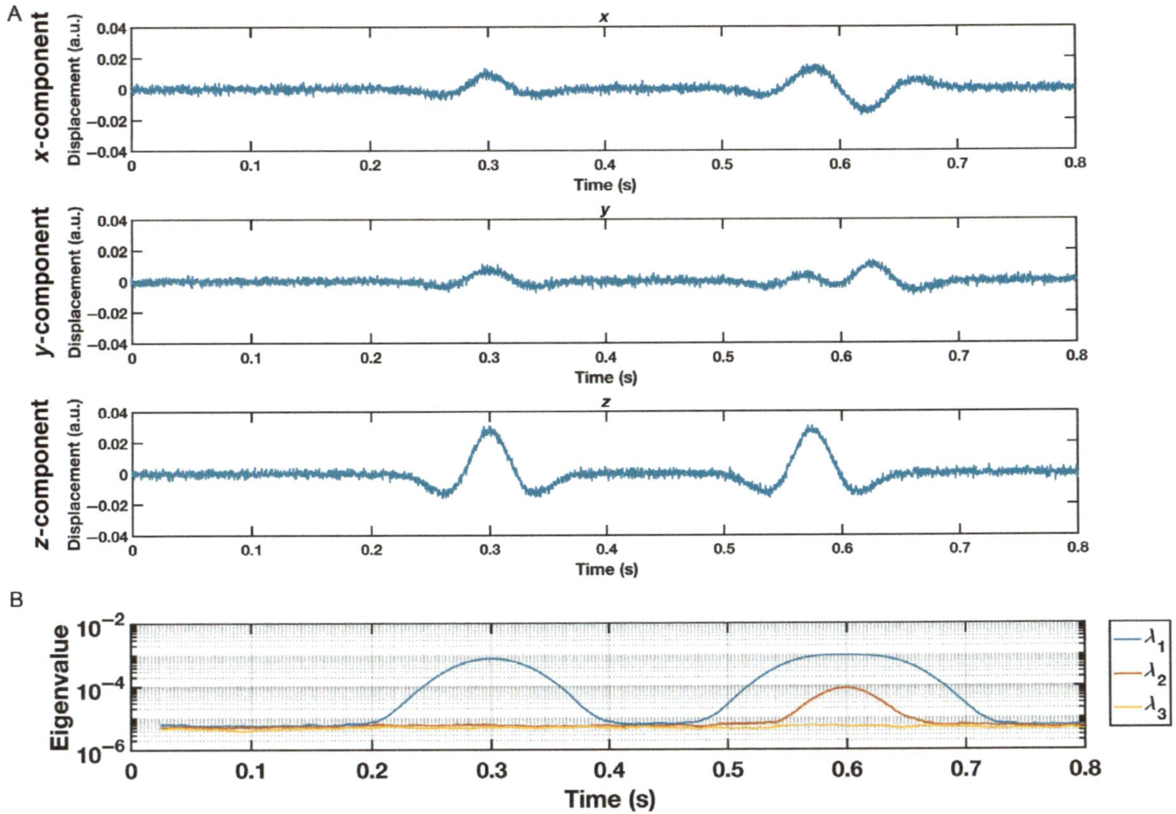

Fig. 7 (A) Same three-component trace set as shown in Fig. 6A. (B) Eigenvalues of the complex covariance matrix computed for the data shown in (A). A 50 ms time window was used when forming the complex covariance matrix around each sample.

restrictions on the types of dataset that may be legitimately processed. The amount of error in direction estimation for overlapping events has been investigated analytically (Greenhalgh et al., 2008).

For a single triaxial station, when coherent noise is present in the analysis window, then the second and third terms of the RHS of Eq. (30) become significant (see Fig. 7). This degrades the benefit of polarization analysis. Similarly, if the SNR falls below unity, the noise terms appearing in Eq. (30) dominate the complex covariance matrix and a successful polarization analysis is not possible. *Single-station polarization data must therefore have a good SNR and must not contain more than one event in each time window used.* In practice, the window for the analysis should optimally be placed to contain all significant samples of the signal (maximizing signal-to-noise energy) without interference from any previous or subsequent events. It should also be as long as possible within these constraints so that the signal may be enhanced above the random noise (Jackson et al., 1991). The choice of window length affects both the detection and resolution of events.

6.3 The analysis domain

When using a complex covariance matrix style of processing, it is important to be clear of the type of signal that is under examination. If it is a plane-polarized event such as a Rayleigh wave, then a complex representation must be used (i.e., the analytic signal in the time domain or the Fourier transform in the frequency domain). However, if the signal is rectilinearly polarized, the analysis may be simplified as the complex covariance matrix reduces to a scaled version of the real covariance matrix.

A rectilinear arrival is easily dealt with in the time domain using the real covariance matrix analysis. This is because the polarization vector being investigated is real (as in Eq. 22). However, if the arrival being examined is a nonrectilinearly polarized event, such as a Rayleigh wave, the signal energy is observed in a plane and its polarization vector is complex (Eq. 25).

6.4 The significance of an eigenvalue

When considering seismic data acquired in the field, it is very important to appreciate the significance of the eigenvalues of the real or complex covariance matrices. Each eigenvalue is proportional to the amount of energy in the principal coordinate axis frame of reference. The eigenvectors provide the directions of these axes within the vector space of the analysis.

Polarization signal processing requires that the number of independent signals in the time window under examination is less than the dimension space of the complex covariance matrix. If this is true, then the complex covariance matrix is singular and this singularity can be used to detect coherent energy.

When there is no random noise, then any nonzero eigenvalue is significant, since it indicates the presence of coherent energy. A single nonzero eigenvalue then represents a purely polarized event. If the complex covariance matrix has two nonzero eigenvalues it indicates two events while three indicates three or more events in the time window. Note that when two or more independent events are recorded by a single three-component station, no useful information can be obtained using polarization analysis. This is because there is no linear dependence apparent in the signal space and the complex covariance matrix is no longer singular. The only option is to work with an array of N triaxial stations, exploiting differential moveout information of the various arrivals as well, thus increasing the dimension of the complex covariance matrix to $3N \times 3N$ (see Rutty & Greenhalgh, 1999).

When noise is present for the single station case, the significance of certain eigenvalues depends on the types of signal and noise present. Let us denote the eigenvalues of the real covariance matrix by λ_i and arrange them in descending order. If there is a rectilinear event in purely random noise (which implies only one event in the time window), then the noise energy within the window may be estimated as $3\lambda_3$ (Kanasewich, 1981; see for example Fig. 7). In this case λ_1 is significant if

$$\lambda_1 \gg \lambda_2 \approx \lambda_3 \tag{32}$$

Again, if several phase-shifted overlapping events with different arrival directions were present within the chosen time window, then even with no random noise, λ_3 may be comparable with λ_1 and so provides no indication to the significance of λ_1.

The eigenvalues computed from an eigendecomposition of the complex covariance matrix derived for a sliding time window of 50 ms for the traces shown in Fig. 6 are displayed in Fig. 7. Whereas the three eigenvalues are small and comparable for the window with purely random noise (see Fig. 6B), either one or two eigenvalues assume significant values for the time windows with one rectilinearly polarized (see Fig. 6C) or two interfering rectilinearly polarized events (see Fig. 6D), respectively.

In summary, the significance of the eigenvalues may usually be tested by comparison with the smallest eigenvalue of the complex covariance matrix. However, this is only true if the signal and any coherent noise within the window are known to span a vector space of smaller dimension than the complex covariance matrix. The smallest eigenvalue is then purely due to random noise.

7. Polarization filtering

The covariance matrix eigendecomposition approach discussed earlier is not only a useful tool in polarization analysis and finding the direction of arrival of an incident wave but it also allows the design of nonlinear polarization filters, similar to the filters that are widely used in optics. For isolated noise-free rectilinear arrivals, all of the energy is expected to be in the direction (eigenvector) corresponding to the largest eigenvalue. In the case of an isolated surface wave, energy will be present in two directions, corresponding to the axes of the polarization ellipse. Ratios of the eigenvalues, as we saw above, provide information on the degree of polarization and the rectilinearity or ellipticity of the signal. This fact can be exploited to build polarization filters that suppress or enhance certain modes.

A very simple polarization filter that enhances signal amplitudes is the rotation of the data into the coordinate frame that is described by its principal axes of motion. This is easily achieved by forming the dot product of the data vector with the three eigenvectors of its covariance matrix:

$$X_i = \mathbf{R_i} \bullet \mathbf{e}_1, \quad Y_i = \mathbf{R_i} \bullet \mathbf{e}_2, \quad Z_i = \mathbf{R_i} \bullet \mathbf{e}_3 \tag{33}$$

where $\mathbf{R_i}$ is the data vector at the ith sample. $\mathbf{e}_1$, $\mathbf{e}_2$, and $\mathbf{e}_3$ are the eigenvectors sorted in descending order of their corresponding eigenvalues The data are represented in its new coordinate frame given by X_i, Y_i, and Z_i.

A common application of this technique is the rotation of the horizontal components into the direction of the source. This can become a necessity if the geophones in the field are not all oriented in the same way. For example, in a VSP the geophones are often lowered down the borehole on a chain. This chain can get twisted so that each geophone is oriented differently in the vertical borehole. The reorientation is achieved by rotating each geophone individually. To do so, a 2×2 covariance matrix has to be formed within a time window around the first arrival. Each sample of the trace is then projected onto the principal eigenvector of this covariance matrix by forming the dot product. In a deviated well the "vertical" component is also not properly aligned. Greenhalgh and Mason (1995) show how to determine the geophone orientation in such a situation.

More complex polarization filters than that described earlier can be built using the polarization parameters given in Section 2.4. They can be used as a gain function to suppress weakly polarized signals and random noise. For example, measures P_4 and B shown in Fig. 9 provide a relatively sharp measure of the degree of polarization that can allow discriminating between polarized signals and random noise.

Vidale (1986) used the analytic signal and the complex covariance matrix to introduce a measure of ellipticity. It is built from the complex eigenvector $\mathbf{e}_1$ corresponding to the largest eigenvalue λ_1. Since the phase in the complex plane of the eigenvectors is initially arbitrary, $\mathbf{e}_1$ has to be rotated by the angle that maximizes the length of its real component. In practice, this is achieved by maximizing function Λ by a search over $\alpha = 0$–180 degrees

$$\Lambda = \sqrt{(\mathrm{Re}\{e_x cis\alpha\})^2 + \left(\mathrm{Re}\{e_y cis\alpha\}\right)^2 + (\mathrm{Re}\{e_z cis\alpha\})^2} \tag{34}$$

where Re{ } is the real part, cisα is $\cos\alpha + j\sin\alpha$ and the components of $\mathbf{e}_1$ are given by e_x, e_y, and e_z. The eigenvector is then rotated by the angle that maximizes Λ. The elliptical component of polarization can then be estimated by

$$P_\mathrm{E} = \frac{\sqrt{1-\Lambda^2}}{\Lambda}. \tag{35}$$

Since the eigenvector is a unit vector, $\sqrt{1-\Lambda^2}$ is the length of the imaginary component of $\mathbf{e}_1$ and thus, P_E is the ratio of the length of the imaginary part to the real part of the eigenvector. P_E is equal to 1 for circularly polarized motion and 0 for linearly polarized motion. If it is used as a gain function, only elliptically polarized events will remain in the signal. On the other hand, it can be used to suppress ground roll by forming a function that minimizes when the elliptical component of polarization is high. Such a function could, for example, look like

$$P_\mathrm{L} = (1 - P_\mathrm{E})^N, \tag{36}$$

where N is a power constant that can be chosen to optimally suppress elliptical motion.

On field data, this kind of ground roll suppression has only limited success because the particle motion of Rayleigh waves often deviates from being purely elliptical. Also, body waves are not always purely linearly polarized. Thus, complete separation between the two is hard to achieve in practice (Kragh & Peardon, 1995). Recent efforts in the suppression of surface waves have therefore been made using not only polarization properties to identify ground roll but also other characteristic attributes. Jin and Shuki (2005) developed a technique that also uses the fact that ground roll is usually rich in low frequencies. To obtain information on the polarization properties, they use a singular value decomposition, an approach that is widely used in multicomponent seismic processing (see Franco & Musacchio, 2001; Jackson et al., 1991). The data matrix can thereby be decomposed into three orthogonal eigenimages by.

$$\mathbf{D} = \mathbf{U}\mathbf{W}\mathbf{V}^{\mathrm{T}} = \sum_{i=1}^{3} \sigma_i \mathbf{u}_i \mathbf{v}_i = \mathbf{E}_1 + \mathbf{E}_2 + \mathbf{E}_3 \tag{37}$$

Here, $\mathbf{D}$ is the data matrix to be decomposed within a specified time window. The vectors $\mathbf{v}_1$, $\mathbf{v}_2$, and $\mathbf{v}_3$ of length three are the columns of matrix $\mathbf{V}$ and represent the eigenvectors of the covariance matrix $\mathbf{D}^{\mathrm{T}}\mathbf{D}$. $\mathbf{u}_1$, $\mathbf{u}_2$, and $\mathbf{u}_3$ of length N (number of samples in the time window) are the columns of the matrix $\mathbf{U}$ representing the eigenvectors of the matrix $\mathbf{D}\mathbf{D}^{\mathrm{T}}$. $\mathbf{W}$ is a 3×3 matrix storing the singular values (σ_1, σ_2, and σ_3) of $\mathbf{D}$ in its diagonal. The singular values are sorted in descending order ($\sigma_1 > \sigma_2 > \sigma_3$). They are the positive square roots of the eigenvalues of the covariance matrix. $\mathbf{E}_1$, $\mathbf{E}_2$, and $\mathbf{E}_3$ are the orthogonal eigenimages or principal components of $\mathbf{D}$.

In order to suppress ground roll, Jin and Shuki (2005) introduce an attribute w that is described by the singular values as

$$w = (\sigma_1 - \sigma_3)(\sigma_2 - \sigma_3) \tag{38}$$

The attribute w is the area of the major principal section of the polarization ellipsoid, adjusted by the minor semiaxis σ_3. An advantage of w is that it includes amplitude information. Ground roll has usually much higher amplitudes than body waves, a fact which helps in the distinction of the two. Using attribute w, filtered data $\mathbf{F}$ can be obtained by

$$\mathbf{F} = (1 - g)\mathbf{D} + g(\mathbf{D} - \mathbf{E}_1 - \mathbf{E}_2) \tag{39}$$

where $\mathbf{D}$ is the original data matrix and $\mathbf{E}_1$ and $\mathbf{E}_2$ are the first two eigenimages of the data matrix after the application of a low pass filter. The quantity g is an integer that indicates if ground roll is present in the time window: $g=1$ if $w > w_g$ and $g=0$ otherwise. Quantity w_g is a threshold value of w above which the ground roll is most likely present. It can be obtained by sorting w in ascending order. w_g can then be identified as the value above which w increases rapidly.

This method uses three characteristic attributes of surface waves to obtain an optimal result: the low-frequency content, the high amplitudes, and the elliptical polarization. Even better signal can be preserved if more than one station is used in the analysis. However, this is only possible if the distance between neighboring stations is short enough to ensure that the ground roll properties do not change much between them.

8. Synthetic examples

The analysis of a synthetic three-component seismogram with the different polarization measures is shown in Figs. 8–10. The seismograms displayed in Fig. 8A contain the arrivals of a P-wave at 200 ms and a Rayleigh wave at 500 ms modeled by convolving a 10-Hz Ricker wavelet with the corresponding polarization vectors given by Eqs. (24) and (25). The arrivals were simulated to have an azimuth of 20 degrees, the P-wave has an incidence angle of 10 degrees, and the Rayleigh wave free parameter ξ was $\pi/8$. Three levels of white Gaussian noise were added to the data with SNRs of 10, 0, and -10 dB. A SNR of 10 dB means that the power of the noise is 10 dB lower than the power of the signal.

The analytic signal was generated for each component of the input data before performing a complex eigenanalysis in the time domain. The time window used when forming the complex covariance matrix around each sample was 50 ms. The eigenvector corresponding to the principal eigenvalue displayed in Fig. 8B determines the principal axis of particle motion. It is plotted as a set of direction cosines DX, DY, and DZ in Fig. 8C and D. Note how it is relatively constant around each picked event. Also note how the eigenvector for the rectilinear event has no imaginary component, whereas that of the plane-polarized event is complex. This is in agreement with Eqs. (24) and (25) which describe the different pure modes of event vibration.

The eigenvalues (Fig. 8B) were then used to produce the picking functions, P_i, plotted in Fig. 9 with the polarization measures P_1, P_2, P_3, and B from Eqs. (3)–(5) and (8), respectively. To compute the polarization measures, we use the values of $\varepsilon = 1$ and $N = 1$ for the Butterworth function (P_2; Eq. 4), and $N = 1$ for P_3 (Eq. 5). All measures maximize when there is a strongly polarized event and minimize when the data are not polarized. The values of P_1, P_2, and P_3 have a maximum value of 1.0. If they attain this value, the motion detected within the window is said to be totally polarized. If the functions reach their minimum value of zero, then the data are totally

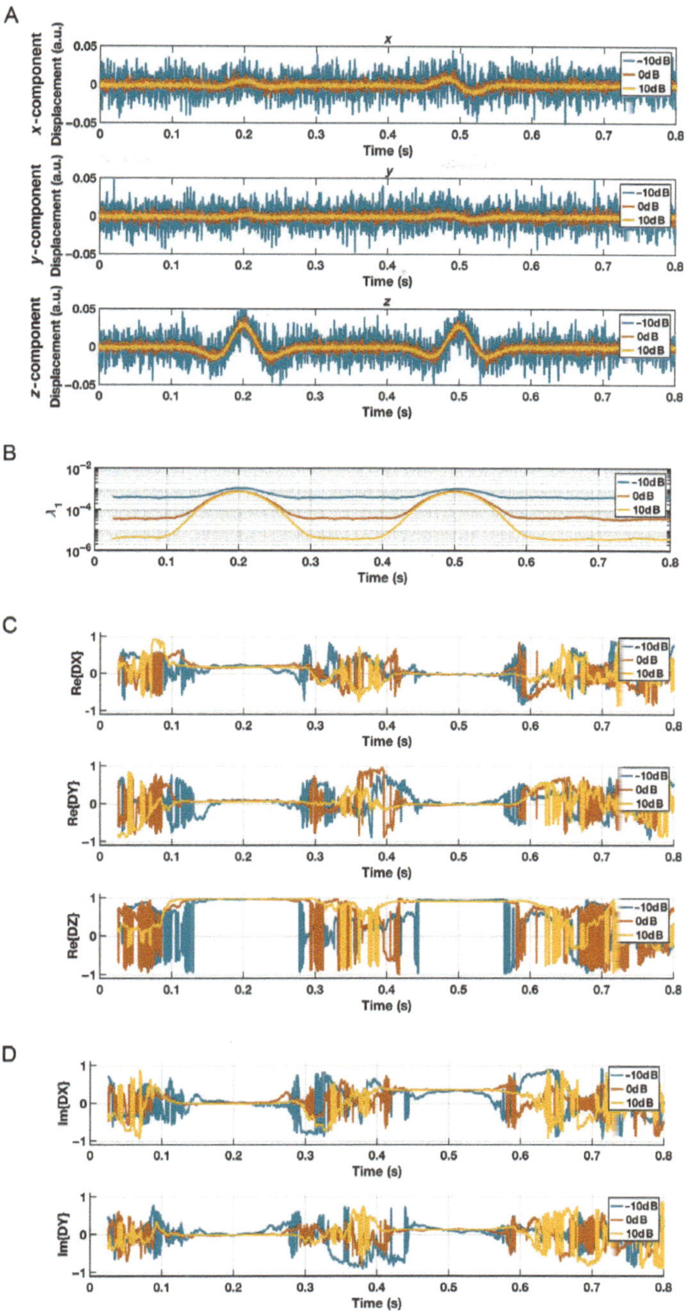

Fig. 8 (A) Three-component seismograms that contain a P-wave at 200 ms and a Rayleigh wave at 500 ms modeled by convolving a 10-Hz Ricker wavelet with the corresponding polarization vectors with an azimuth of 20 degrees, a P-wave incidence angle of 10 degrees, and a Rayleigh wave ellipticity of $\pi/8$. Three levels of white Gaussian noise were added to the data with SNRs of 10, 0, and -10 dB. (B) Principal eigenvalue λ_1 obtained from an eigendecomposition. A 50 ms time window was used when forming the complex covariance matrix around each sample of the traces shown in (A). (C) Real and (D) imaginary part of the three components of the principal eigenvector (direction cosines DX, DY, and DZ). Note that the imaginary part of DZ is zero and therefore not shown.

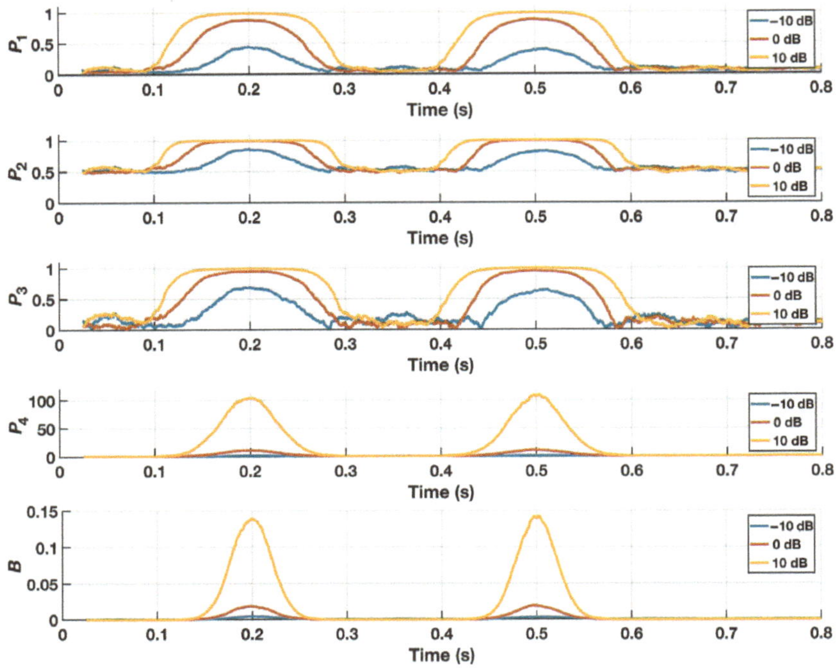

Fig. 9 Picking functions P_1, P_2, P_3, P_4, and B from Eqs. (3)–(6) and (8), respectively, computed for the data shown in Fig. 8A. To compute the polarization measures, we use the values of $\varepsilon=1$ and $N=1$ for the Butterworth function (P_2; Eq. 4), and $N=1$ for P_3 (Eq. 5).

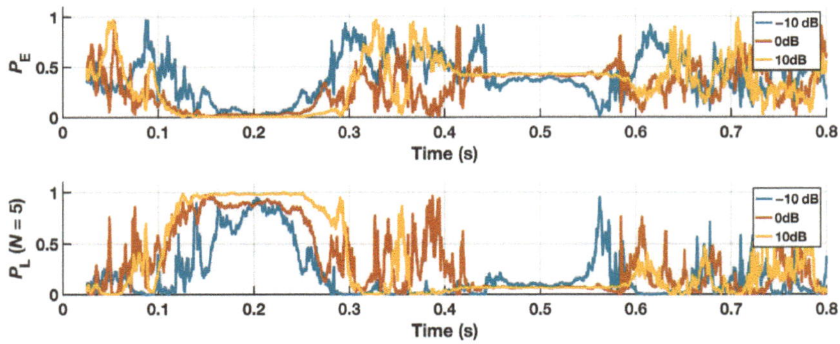

Fig. 10 Measure of ellipticity P_E (Eq. 35) and gain function to suppress elliptically polarized events P_L (computed with Eq. 36 and $N=5$). Inputs are the three-component seismograms shown in Fig. 8A.

unpolarized. Any value between these values indicates that the seismic event is partially polarized. Note that a purely elliptic signal is no less polarized than a pure rectilinear signal, there is just a phase difference between components. They both correlate equally well when working with the complex analytic signal (coherency matrix) and yield a single nonzero eigenvalue λ_1. There is no upper bound on the function P_4. This function maximizes when the minor eigenvalues in the denominator tend to zero, similar to B. The measures P_4 and B produce sharper peaks than the other functions.

Both events at 200 and 500 ms are clearly detected in the picking functions, which reach their maxima where the signals occur. Note that the width and the amplitude of all picking functions decrease with increasing noise level.

The measure of elliptical motion P_E (Eq. 35) is displayed for the same synthetic data example as discussed earlier in Fig. 10. P_E passes the Rayleigh wave at 500 ms reaching a value of around 0.4, which closely corresponds to the true Rayleigh wave free parameter value ξ of $\pi/8$, but is insensitive to the P-wave at a time of 200 ms. The function P_L is created using Eq. (35) and a power constant of $N=5$ could be used to suppress the Rayleigh wave.

9. Real data examples

The conditions in the field often deviate considerably from some of the assumptions that were made in the synthetic data analysis discussed in the previous sections. Thus, it is important to test the applicability of the proposed algorithms on field data. In this section, we give examples of single-station polarization processing on real datasets.

9.1 Two-component rotation to orient borehole geophones

The incoherent events observed in the raw horizontal components of the borehole-data example shown in Fig. 11 demonstrate the necessity of applying a two-component rotation to the horizontal components of VSP datasets. This VSP dataset was recorded during a field campaign in northern Switzerland with the goal to image the topmost 1 km depth around the Benken borehole (Nagra, 2001). The seismograms of the two horizontal components are shown in the top and the middle panels of Fig. 11. The principal axis of motion was determined in a 10-ms time window around the first arrivals for each trace individually using the covariance matrix eigendecomposition technique. The data vector was then projected onto this axis (see Eq. 33) to achieve a reorientation of the sensors into the direction of the first arrivals. If straight paths between the source and the receivers are assumed this is also the direction of the source. The result of rotating each trace into the principal axis of motion is shown in the bottom panel of Fig. 11.

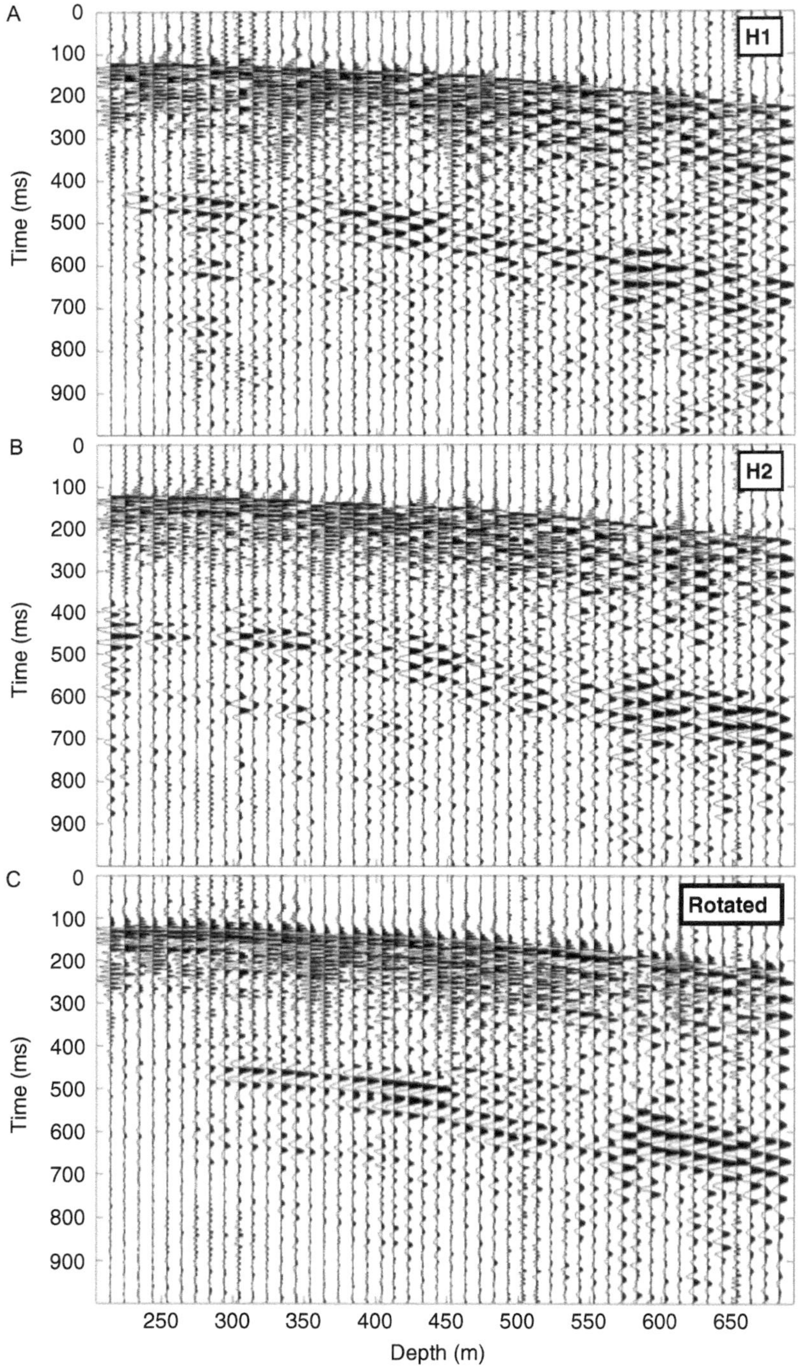

Fig. 11 Two horizontal components recordings *H*1 and *H*2 measured during a VSP experiment (*top and middle*) and after rotation into the direction of principal motion of the first arrivals.

After the rotation, the channel-to-channel correlation of the arrivals is drastically increased. The reason for this improvement is that the geophones in the borehole were not all oriented in the same way.

9.2 Analysis of the M 6.8 Jan Mayen Earthquake

We will demonstrate the application of the polarization analysis using the three-component recordings of the M 6.8 Jan Mayen earthquake from 30 August 2012 (71.441°N 10.605°W; 14 km depth; 13:43:25.170 UTC) measured by the Black Forrest Observatory station, Germany (GE.BFO). Fig. 12 displays the seismic traces plotted with time relative to the event origin time. The ray-theoretical arrival times of a number of key phases were computed using the PREM model (Crotwell, Owens, & Ritsema, 1999; TauPyModel tool of ObsPy, Krischer et al., 2015). The arrival times of the P-, S-, Rayleigh waves that travel through the mantle (denoted by R; crustal Rayleigh waves are marked by Rg), and Love waves are marked with colored lines in Fig. 12.

We used a 5-s time window when forming the complex covariance matrix around each sample. The three eigenvalues resulting from the eigendecomposition are shown as a function of time in Fig. 13. For the time interval before the P-wave first arrival observed at around 321 s, the three eigenvalues are close to each other but not identical, suggesting that this time window of the traces is dominated by polarized microseismic noise (e.g., Schimmel, Stutzmann, Ardhuin, & Gallart, 2011). The arrival of the P- and S-waves lead to an increase of all three eigenvalues and the distinct principal eigenvalue indicates polarized events. The most pronounced increase associated with the largest spread of all eigenvalues is observed after the arrival of the surface waves indicating that these events are the most polarized.

The polarization attributes P_1 to P_4 and B, computed with $\varepsilon = 100$ and $N = 4$ for P_2 and $N = 0.5$ for P_3, are displayed in Fig. 13. Measures P_1 and P_3 reach values close to 1 for the P-wave arrival and the surface wavetrain, whereas P_2 shows a clear spike for the P-wave first arrival and the sP-arrival following immediately afterward as well as a constant high level for the Love-wavetrain. Attributes P_4 and B provide measures of polarization and allow distinguishing polarized signal from noise. Note that B increases by about three orders of magnitude with the P-wave first arrival. Both P4 and B indicate the continuing arrival of polarized events at 2000 s even though the amplitudes in the original recordings in Fig. 12 have already significantly decayed.

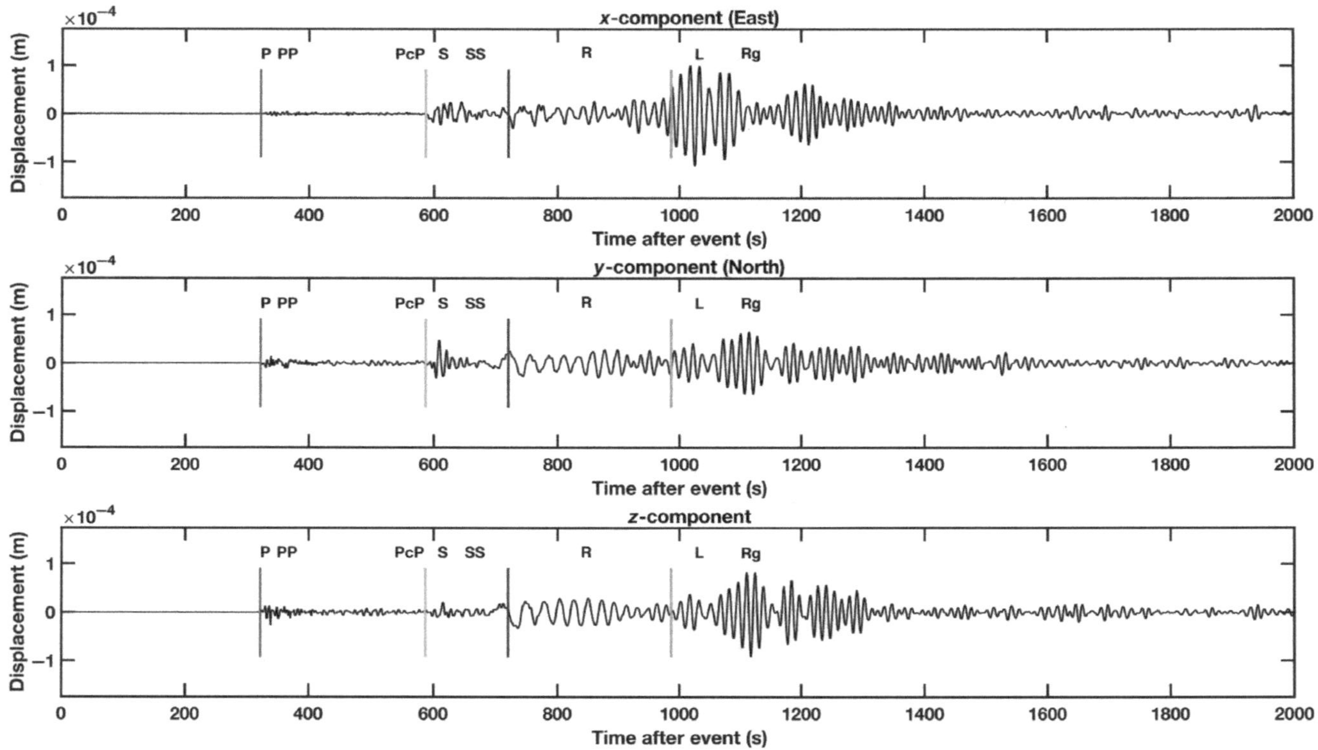

Fig. 12 Three-component recording of the M 6.9 earthquake in Jan Mayen on August 30, 2012 by the station BFO. Names of the most important phases are given at their corresponding arrival times and *colored lines* denote onsets of a few key phases.

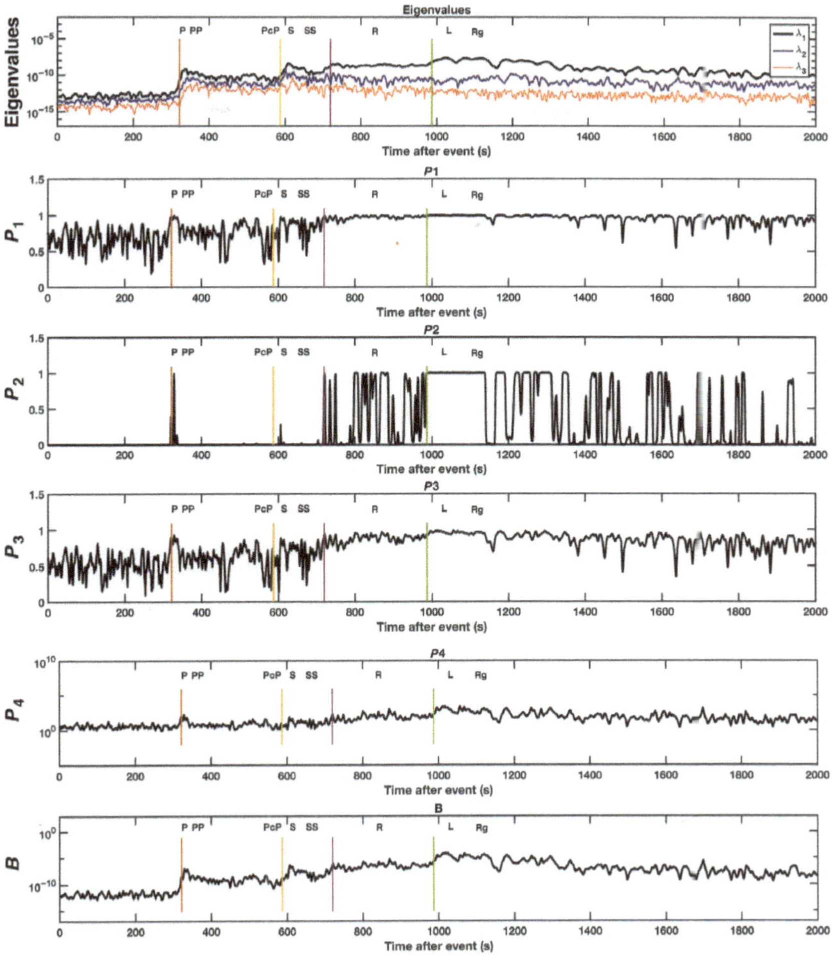

Fig. 13 Eigenvalues as well as polarization attributes P_1 to P_4 and B (Eqs. 3–6 and 8) computed for the data shown in Fig. 12. A 5-s long time window was used when forming the complex covariance matrix around each sample. Names of the most important phases are given at their corresponding arrival times and *colored lines* denote onsets of a few key phases.

The ellipticity attribute P_E exhibits strong fluctuations from sample to sample (black line in Fig. 14A). The overall trend becomes clearer when smoothing the attribute with a 101-s long running mean (red line in Fig. 14A). The smoothed P_E values are also displayed as colored background with red corresponding to primarily linear polarization and blue indicating primarily elliptical polarization in all panels of Fig. 14. Clearly, the P-wave first arrival and following wavetrain until the arrival of the first S-wave arrival are the most

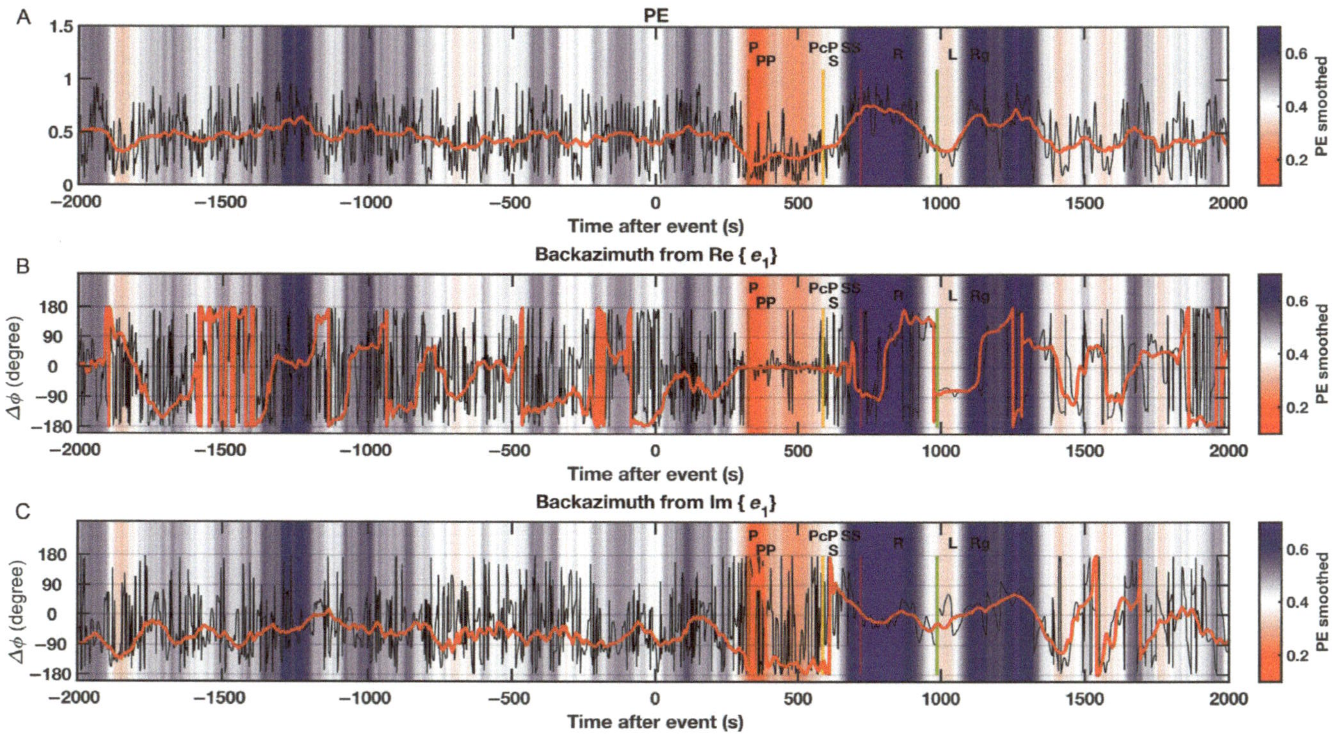

Fig. 14 (A) Ellipticity measure P_E, (B) back-azimuth estimated from the real part and (C) imaginary part of the principal eigenvector shown for a time window ranging from 2000 s before the event to 2000 s after the event (see Fig. 12 for data). The estimated back-azimuth is displayed as deviation from the nominal back-azimuth of 345.4 degrees. A 5-s long time window was used when forming the complex covariance matrix around each sample and the *black lines* show the sample-by-sample estimates. The *red lines* display the 101-s running mean of PE (in A), and the back-azimuth estimated from the smoothed real part (in B) and imaginary part (in C) of the principal eigenvector (101-s running mean). The *background color* corresponds to the smoothed P_E values for all three plots. Names of the most important phases are given at their corresponding arrival times and *colored lines* denote onsets of a few key phases.

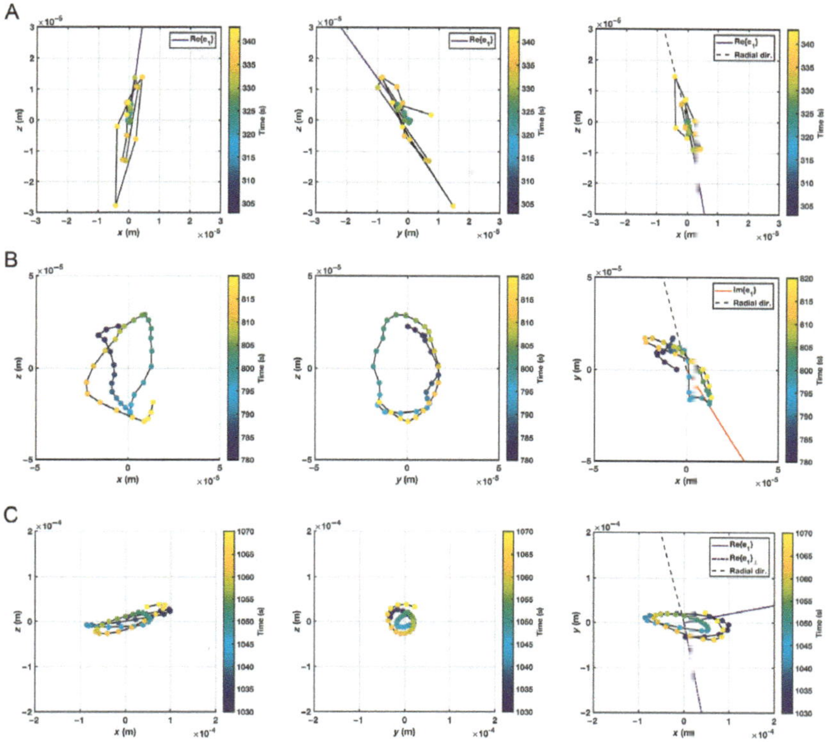

Fig. 15 Hodograms for 40-s long time windows extracted from the data shown in Fig. 12 for (A) the P-wave first arrival at 321 s, (B) the mantle Rayleigh waves at 800 s, and (C) the Love waves at 1050 s. The *color of the dots* corresponds to the sample time to give an impression of the time evolution (see *color bar* for times). The *dashed lines* in the *x–y* plane hodograms show the nominal radial direction (back-azimuth of 345.4 degrees), *blue and red lines* correspond to the projection of the real and imaginary parts of the principal eigenvector onto the *x–y* plane, respectively. In (C), the direction perpendicular to the principal eigenvector is shown as *dash-dotted line* as well.

linearly polarized events, whereas the two Rayleigh wavetrains R and Rg are the most elliptically polarized events observed in the recording. The inspection of hodograms displayed in Fig. 15A and B confirms the rectilinearly polarized nature of the P-wave first arrival and the elliptical polarization of the R phases, respectively. The R and Rg wavetrains are separated by a short time window dominated by the linearly polarized Love waves at around 1000 s. The hodograms shown in Fig. 15C also illustrate the linear polarization in the horizontal plane of the Love waves.

The principal eigenvector can be used to compute the azimuth ϕ and the inclination angle θ for each sample (see Fig. 3). As a consequence of the eigenanalysis of the complex covariance matrix (Eq. 27), the eigenvectors are complex. However, note that the imaginary part of the vertical component of the principal eigenvector is always equal to zero. Real and imaginary contributions result from the rectilinearly and elliptically polarized motions, respectively. Therefore, we analyze the real part of the principal eigenvector to determine ϕ and θ for the rectilinearly polarized P-wave first arrival, which are also plotted in Fig. 15A. The resultant apparent P-wave incidence angle ϕ_{app} and θ are 37.6 and 169.4 degrees, respectively. In comparison, the true incidence angle of the P-wave first arrival determined for the PREM model is 28.4 degrees. The extracted θ value for the P-wave corresponds to a (receiver-to-source) back-azimuth of 349.4 degrees, which is very close to the reported back-azimuth of 345.4 degrees (see Fig. 15A).

A sample-by-sample back-azimuth analysis for both the real and imaginary parts of the principal eigenvector is displayed in Fig. 14B and C (black lines) together with back-azimuth estimates obtained by smoothing the eigenvector components with a 101-s long running mean (red lines). Considering the smoothed P_E attribute values shown as a background color in Fig. 14, we can interpret the real eigenvector components to provide back-azimuth estimates for the rectilinearly polarized events with small P_E values (red color) such as the P- and Love waves (Fig. 14B). Blue colors mark elliptically polarized events such as the Rayleigh waves, for which the imaginary parts provide interpretable back-azimuth values (Fig. 14C). We plotted all back-azimuth estimates as deviations from the reported back-azimuth of 345.4 degrees.

For the R waves shown in Fig. 14B, back-azimuth deviates by about 22 degrees from the expected radial direction, which could be due to, for example, interfering waves. The Love waves observed at 1050 s exhibit a back-azimuth close to the expected value, when considering that Love waves are rectilinearly polarized but at 90 degrees to the propagation direction. In summary, consistent back-azimuth estimates could be extracted for both the P- and surface waves. Interestingly enough, the time period before the P-wave first arrival, which was previously observed to be primarily elliptically polarized, is characterized by a quite consistent back-azimuth of around 300 degrees, which corresponds to an arrival direction from the Atlantic, a common source region of microseismic noise.

The data example presented in Fig. 12 is characterized by a high SNR. To explore the performance of the eigenanalysis for increased noise levels, random Gaussian white noise of different levels was added to the observed three-component data (Fig. 16). We then computed the complex

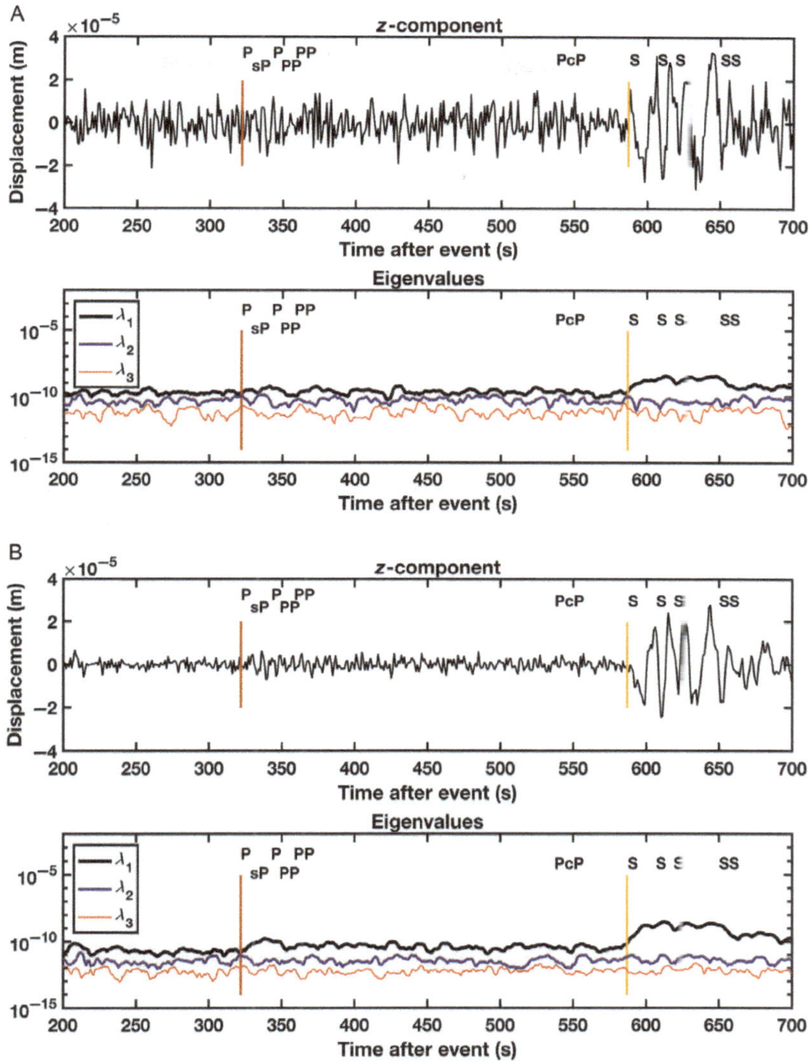

Fig. 16 Artificially noise-contaminated recordings of the M 6.9 earthquake in Jan Mayen on 30. August 2012 recorded by station BFO (see Fig. 12) and corresponding eigenvalues for a noise of power (A) 5 dB and (B) 15 dB added.

covariance matrix and corresponding eigenvalues for each sample using the same 5-s long time window. We deemed a visible increase in the first eigenvalue relative to the other eigenvalues as an indication that the analysis presented earlier can successfully be carried out (see Eq. 32). When adding noise leading to a SNR of 5 dB (Fig. 16A) and 15 dB (Fig. 16B) in power (compared to the power of the uncontaminated data series), then the S- and P-wave arrivals, respectively, are associated with a distinct increase of the first eigenvalue.

9.3 Surface reflection survey in Switzerland

This example is intended to demonstrate the application of ground roll attenuation using the polarization filter approach of Jin and Shuki (2005) discussed in Section 7. The data displayed in the left column of Fig. 17 is just one shot gather from many recorded in a surface seismic reflection survey carried out in northern Switzerland (Madritsch et al., 2013). An explosive

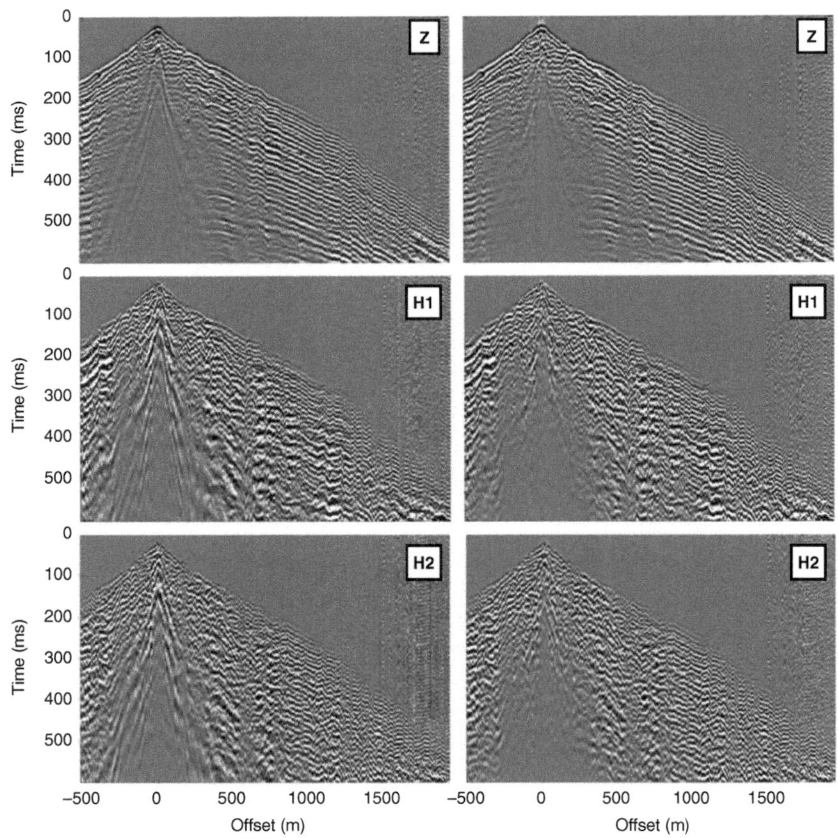

Fig. 17 Ground roll attenuation achieved by a nonlinear polarization filter using singular value decomposition of the complex covariance matrix along a sliding time window for each trace in a shot gather. (*Left*) Input data of the vertical (*Z*) and horizontal (*H1*, *H2*) components. (*Right*) Output of the polarization filter algorithm.

source was used to acquire the displayed source gather. Unfortunately, the data quality on the two horizontal components *H*1 and *H*2 is not as good as that on the vertical component *Z*. Polarization analysis was carried out on one triaxial trace at a time and succeeded in identifying most of the elliptically polarized energy due to Rayleigh waves (observed to have the steep moveout on the shot gathers). The nonlinear polarization filter has removed most of the surface wave contamination of the total signal, as shown in the right column of Fig. 17. The reflection signals are preserved in the data.

10. Discussion

Seismic polarization analysis and direction finding are prone to errors caused by physical effects and practical aspects. This has to be kept in mind during the interpretation of such multicomponent data. In the following section, we briefly describe some of the most important problems that can cause the polarization analysis to fail.

10.1 Experimental difficulties of single-station seismic direction finding

10.1.1 Random and coherent noise effects

In the presence of purely random noise (i.e., it is equal in energy on all channels), the eigenanalysis of the covariance matrix theoretically always yields the correct direction estimate, regardless of the SNR (Greenhalgh et al., 2008). In practice, this is not the case because the direction finding algorithms all operate within a finite time window. In this window, noise will not cancel out completely. This would only happen if the time window were infinitely long. In a finite time window, the direction estimate will always have an error related to the variance of the noise distribution. Increasing the window length can minimize this error. However, our experience is that for moderate to high-quality data, keeping the analysis time window short to avoid multiple events within one time window (coherent noise) is more critical than increasing the time window to reduce the impact of random noise.

Multipathing can cause events to overlap within a time window, resulting in an error in the direction estimation. This error has been quantified by Greenhalgh et al. (2008) for two interfering transient wavelets. They show the effects of relative amplitude, arrival angle, and the time delay of the two wavelets.

10.1.2 Calibration errors

The fact that gains (geophone transduction factors, ground coupling variations, and/or preamplifier levels on the recording system) are not always constant across the three sensor channels can lead to errors in the estimation of the polarization. For the biaxial problem, these errors can be quantified with a simple calculation. Let us consider two orthogonal channels X and Z. X has unit gain and Z has an unknown gain G causing a distortion in the direction estimation. The apparent direction θ_A and the true direction θ_T can be described by:

$$\tan\theta_A = \frac{Z}{X} = G\frac{dz}{dx} \tag{40}$$

and

$$\tan\theta_T = \frac{dz}{dx}. \tag{41}$$

Here, dz and dx represent the relative amplitudes in the case of constant gain on both channels. The gain misalignment introduces an error $\delta\theta$ between the true direction and the apparent direction:

$$\delta\theta = \theta_T - \theta_A = \tan^{-1}\left(\frac{dz}{dx}\right) - \tan^{-1}\left(G\frac{dz}{dx}\right) \tag{42}$$

G is defined in decibels by:

$$G = 20\log\left(\frac{Z}{dz}\right) \tag{43}$$

The error can now be plotted against the true polarization direction for different gains on channel Z. Fig. 18 shows results of this analysis and makes it clear that polarization analysis fails if the difference in sensitivity/gain/coupling between the channels is large. To prevent errors caused by mis-wiring or poor mechanical coupling of the sensors, it is essential to extensively test (or perform calibration) in the field by firing calibration shots in different octants. These errors have to be accounted for in the analysis.

11. Conclusions and outlook

Given a specific time window over data from a single triaxial station, the polarization properties of the particle motion within this window can be estimated for event characterization (e.g., mode and arrival direction) as well

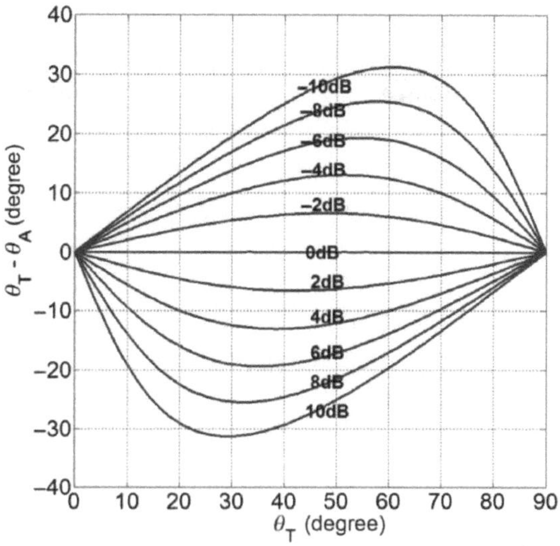

Fig. 18 Quantified error in the direction estimation caused by gain misalignments between the horizontal and the vertical channel. The gains are applied to the vertical channel only, unit gain is assumed for the horizontal channel.

as wavefield filtering. For example, a filter operator can be calculated to enhance rectilinearly polarized signals using the relative size of the eigenvalues. This function may then be applied to the data and is the basis for polarization filtering. Furthermore, any window of the original three-component seismogram can be rotated into its principal axes of motion given by the eigenvectors of its complex covariance matrix. To obtain information about the whole trace, the analysis window can be placed at different positions along the data, or repeated in overlapping sliding windows.

The limitations of the polarization technique must be understood before applying them to seismic data. The main requirements for the data are: (1) as large a SNR as possible and (2) no coherent noise in the analysis window. The second requirement is the more demanding to meet for seismic data, since there are normally several events present on the seismic section. If the wavelets are sufficiently separated, then the time window chosen for the analysis should be less than the event separation.

If multiple events interfere (i.e., the smallest eigenvalue of the covariance matrix is not purely due to random noise) and multiple station data are available, it may be worthwhile expanding the dimension of the covariance matrix to use more stations (De Meersman and Kendall, 2005; Rutty & Greenhalgh, 1999). This also provides an increase in SNR if the signal is coherent across the spatial array.

Alternatively, the dimension of the covariance matrix can be expanded by including new observables into the analysis. With the recent advent of sensitive rotational motion sensors, direct observations of ground rotational motions are now possible (e.g., Schmelzbach et al., 2018). Recently, Sollberger, Greenhalgh, Schmelzbach, Van Renterghem, and Robertsson (2018) demonstrated how rotational motion recordings can be used to expand the covariance matrix dimension for single-station polarization analysis. They showed that polarization analysis using a single six-component station (recording three components of rotational motion in addition to the conventional three components of translational motion) provides similar capabilities as conventional array analysis techniques (e.g., *f–k* analysis) when it comes to characterizing the seismic waves in terms of their propagation direction, velocity, and wave type. Additionally, due to the expanded dimension of the covariance matrix, such six-component polarization techniques also succeed when two events are overlapping in the analysis time window. This can be exploited to devise polarization filters, for example, for surface wave suppression, that cause significantly less damage to the reflection signal (Sollberger, Schmelzbach, Van Rengerghem, Greenhalgh, & Robertsson, 2017). These applications demonstrate that polarization analysis can be extended beyond the conventional analysis of translational motion recordings.

Single-station polarization analysis may gain increased importance in cases where only one triaxial station can be installed such as for planetary seismology missions. For example, the InSight mission to Mars is centered around a single seismometer to study the interior of Mars, hence, will require the extensive use of various single-station multicomponent analysis techniques (e.g., Panning et al., 2015).

Acknowledgments

The data of station GE.BFO are part of the GEOFON network (GEOFON Data Centre, 1993; doi: 10.14470/TR560404). We thank Simon Staehler for his assistance in preparing the Jan Mayen data example and Pascal Edme for a thorough review of an earlier version of this chapter. Furthermore, we thank Nagra (Nationale Genossenschaft für die Lagerung radioaktiver Abfälle) for proving access to the surface seismic reflection and VSP dataset.

Appendix: Amplitudes of reflected waves at a free surface

The polarization templates for displacement recordings of body waves at the free surface given in Eq. (24) depend on the amplitudes of the incident and reflected waves (see Fig. 4). In the case of an incident P-wave at the free surface with amplitude A_P, the amplitudes of the downward reflected P-wave, A_{PP}, and the downward reflected P-to-S conversion, A_{PS}, normalized by the incident wave amplitude A_P are (Achenbach, 1973; see also Fig. 4A):

$$\frac{A_{PP}}{A_P} = \frac{\sin 2\theta_P \sin 2\theta_S - \kappa^2 \cos^2 2\theta_S}{\sin 2\theta_P \sin 2\theta_S + \kappa^2 \cos^2 2\theta_S} \tag{A1}$$

and

$$\frac{A_{PS}}{A_P} = \frac{2\kappa \sin 2\theta_P \cos 2\theta_S}{\sin 2\theta_P \sin 2\theta_S + \kappa^2 \cos^2 2\theta_S} \tag{A2}$$

with κ marking the P-to-S-wave velocity ratio, and θ_P and θ_S being the P- and S-wave incidence angle (note: $\sin\theta_S = \kappa^{-1} \sin\theta_P$).

For an S-wave impinging at the free surface with amplitude A_S, the amplitudes of the downward reflected S-wave, A_{SS}, and the downward reflected S-to-P conversion, A_{SP}, normalized by the incident wave amplitude A_S are (Achenbach, 1973; see also Fig. 4B):

$$\frac{A_{SS}}{A_S} = \frac{\sin 2\theta_S \sin 2\theta_P - \kappa^2 \cos^2 2\theta_P}{\sin 2\theta_S \sin 2\theta_P + \kappa^2 \cos^2 2\theta_S} \tag{A3}$$

and

$$\frac{A_{SP}}{A_S} = \frac{-\kappa \sin 4\theta_S}{\sin 2\theta_S \sin 2\theta_P + \kappa^2 \cos^2 2\theta_S}. \tag{A4}$$

For a more extensive treatment of reflections of plane P- and S-waves at the free surface, the interested reader is referred to Aki and Richards (2002, 130–139).

References

Achenbach, J. D. (1973). *Wave propagation in elastic solids*. Vol. 16. Elsevier Science Publishers B.V.

Aki, K., & Richards, P. G. (2002). *Quantitative seismology* (2nd ed.). University Science Books.

Bataille, K., & Chiu, J. M. (1991). Polarization analysis of high-frequency, three-component seismic data. *Bulletin of the Seismological Society of America, 81*, 622–642.

Benhama, A., Cliet, C., & Dubeset, M. (1988). Study and applications of spatial directional filtering in three component recording. *Geophysical Prospecting, 36*, 591–613.

Born, M., & Wolf, E. (1975). *Principles of optics*. Oxford: Pergamon Press.

Christoffserson, A., Husebye, E. S., & Ingate, S. F. (1988). Wavefield decomposition using ML-probabilities in modeling single site 3-component records. *Geophysical Journal, 93*, 197–213.

Cichowicz, A., Green, R. W. E., & Brink, A. v.-Z. (1988). Coda polarization properties of high-frequency microseismic events. *Bulletin of the Seismological Society of America, 78*, 1297–1318.

Cliet, C., & Dubesset, M. (1987). *Polarization analysis in three-component seismcis: Application to well seismic profiles*. Institut Francais de Petrole. Internal Report 35240.

Crotwell, H. P., Owens, T. J., & Ritsema, J. (1999). The TauP toolkit: Flexible seismic travel-time and ray-path utilities. *Seismological Research Letters, 70*, 154–160.

De Meersman, K., & Kendall, R. (2005). A complex SVD-polarization filter for ground roll attenuation on multi-component data. Z-99. In *67th Mtg. European association of geoscientists and engineers, Madrid*.

De Meersman, K., Kendall, J. M., & Van der Baan, M. (2009). The 1998 Valhall microseismicity: An integrated study of relocated sources, seismic multiples and S-wave splitting. *Geophysics, 74*(5), B183–B195.

Esmersoy, C. (1984). Polarization analysis, orientation and velocity estimation in three-component VSP. In M. N. Toksoz & R. R. Stewart (Eds.), *Vertical seismic profiling part B: Advanced concepts* (pp. 236–255). London: Geophysical Press.

Flinn, E. A. (1965). Signal analysis using rectilinearity and direction of particle motion. *Proceedings of the IEEE, 53*, 1874–1876.

Franco, R., & Musacchio, G. (2001). Polarization filter with singular value decomposition. *Geophysics, 66*, 932–938.

Gaiser, J. (2016). 3C seismic and VSP: Converted waves and vector wavefield applications. In Society of Exploration Geophysicists.

Gal'perin, E. I. (1984). *The polarization method of seismic exploration*. Reidel Publishing.

Greenhalgh, S., & Mason, I. M. (1995). Orientation of a downhole triaxial geophone. *Geophysics, 60*, 1234–1237. https://doi.org/10.1190/1.1443852.

Greenhalgh, S. A., Mason, I. M., Lucas, E., Pant, D., & Eames, R. T. (1990). Controlled direction reception filtering of P- and S-waves in τ-p space. *Geophysical Journal International, 100*, 221–234.

Greenhalgh, S. A., Mason, I. M., & Zhou, B. (2005). An analytical treatment of single station triaxial seismic direction finding. *Journal of Geophysics and Engineering, 2*(8–15), 2005.

Greenhalgh, S. A., Zhou, B., & Rutty, M. (2008). Effect of coherent noise on single-station direction of arrival estimation. *Journal of Seismology, 12*, 377–385.

Harris, D. B. (1990). Comparison of the directional estimation performance of high-frequency seismic arrays and three-component stations. *Bulletin of the Seismological Society of America, 80*, 1951–1968.

Hearn, S., & Hendrick, N. (1998). A review of single-station time-domain polarization analysis techniques. *Journal of Seismic Exploration, 8*, 181–202.

Jackson, G. M., Mason, I. M., & Greenhalgh, S. A. (1991). Principal component transforms of triaxial recordings by singular value decomposition. *Geophysics, 56*, 528–533.

Jarpe, S., & Dowla, F. (1991). Performance of high-frequency three-component stations for azimuth estimation from regional seismic phases. *Bulletin of the Seismological Society of America, 81*, 987–999.

Jin, S., & Shuki, R. (2005). Ground roll detection and attenuation by 3C polarization analysis. B020. In *67th Mtg. European association of geoscientists and engineers, Madrid*.

Jurkevics, A. (1988). Polarization analysis of three-component array data. *Bulletin of the Seismological Society of America, 78*, 1725–1743.

Kanasewich, E. R. (1981). *Time sequence analysis in geophysics*. University of Alberta Press.

Knowlton, K. B., & Spencer, T. W. (1996). Polarization measurement uncertainty on three-component VSPs. *Geophysics, 61*, 594–599.

Kragh, E., & Peardon, L. (1995). Ground roll and polarization. *First Break, 13*, 369–378.

Krischer, L., Megies, T., Barsch, R., Beyreuther, M., Lecocq, T., Caudron, C., et al. (2015). ObsPy: A bridge for seismology into the scientific Python ecosystem. *Computational Science & Discovery, 8*(1), 14003.

Kulesh, M., Diallo, M. S., Hoschneider, M., Kurennaya, K., Krüger, F., Ohrnberger, M., et al. (2007). Polarization analysis in the wavelet domain based on the adaptive covariance method. *Geophysical Journal International, 170*, 667–678.

Madritsch, H., Meier, B., Kuhn, P., Roth, P., Zingg, O., Heuberger, S., et al. (2013). *Regionale strukturgeologische Zeitinterpretation der Nagra 2D-Seismik 2011/12, NAB 13-10 (in German).*

Magotra, N., Ahmed, N., & Chael, E. (1987). Seismic event detection and source location using single-station (three-component) data. *Bulletin of the Seismological Society of America, 77*, 958–971.

Magotra, N., Ahmed, N., & Chael, E. (1989). Seismic event detection and source location. *IEEE Transactions on Geoscience and Remote Sensing, 27*, 15–23.

Montalbetti, J. F., & Kanasewich, E. R. (1970). Enhancement of teleseismic body phases with polarization filter. *Geophysical Journal of the Royal Astronomical Society, 21*, 119–129.

Nagra. (2001). *Sondierbohrung Benken Untersuchungsbericht, NTB 00-01 (in German).*

Nuttli, O. (1961). The effect of the Earth's surface on the S wave particle motion. *Bulletin of the Seismological Society of America, 51*, 237–246.

Panning, M. P., Beucler, É., Drilleau, M., Mocquet, A., Lognonné, P., & Banerdt, W. B. (2015). Verifying single-station seismic approaches using Earth-based data: Preparation for data return from the InSight mission to Mars. *Icarus, 248*, 230–242.

Rene, R. M., Filter, J. L., Forsyth, P. M., Kim, K. Y., Murray, D. J., Waters, J. K., et al. (1986). Multi-component seismic studies using complex trace analysis. *Geophysics, 51*, 308–325.

Rutty, M. J., & Greenhalgh, S. A. (1999). Correlation using triaxial data from multiple stations in the presence of coherent noise. In R. L. Kirlin & W. Done (Eds.), *Covariance analysis for seismic signal processing* (pp. 291–322). Geophysical Developments No. 8, Society of Exploration Geophysics.

Samson, J. C. (1983). Pure states, polarized waves and principal components in the spectra of multiple, geophysical time series. *Geophysical Journal of the Royal Astronomical Society, 72*, 647–664.

Schimmel, M., Stutzmann, E., Ardhuin, F., & Gallart, J. (2011). Polarized Earth's ambient microseismic noise. *Geochemistry, Geophysics, Geosystems, 12*, Q07014.

Schmelzbach, C., Donner, S., Igel, H., Sollberger, D., Taufiqurrahman, T., Bernauer, F., et al. (2018). Advances in 6-C seismology: Applications of combined translational and rotational measurements in global and exploration seismology. *Geophysics, 83*, WC53–WC69.

Shieh, C. F., & Herrmann, R. B. (1990). Ground roll rejection using polarization filters. *Geophysics, 55*, 1216–1222.

Sollberger, D., Greenhalgh, S. A., Schmelzbach, C., Van Renterghem, C., & Robertsson, J. O. A. (2018). 6-C polarization analysis using point-measurements of translational and rotational ground-motion: Theory and applications. *Geophysical Journal International, 213*, 77–97.

Sollberger, D., Schmelzbach, C., Van Rengerghem, C., Greenhalgh, S. A., & Robertsson, J. O. A. (2017). Automated, six-component, single-station ground-roll identification and suppression by combined processing of translational and rotational ground-motion. In *SEG Annual Meeting, Houston, TX, USA.*

Sriram, R., DiSiena, J., Mack, H., Tatham, B., Wiggins, W., Shah, P., et al. (1990). Recording and processing vector wave field data: A review of the a989 SEG summer research workshop. *The Leading Edge, 9*, 11–16.

Vidale, J. E. (1986). Complex polarization analysis of particle motion. *Bulletin of the Seismological Society of America*, *76*, 1393–1405.

Winterstein, D. F., & Meadows, M. A. (1991). Shear-wave polarizations and subsurface stress directions at lost hills field. *Geophysics*, *56*, 1331–1348.

Further reading

Rutty, M. J., & Greenhalgh, S. A. (1993). The correlation of seismic events in multicomponent data in the presence of coherent noise. *Geophysical Journal International*, *113*, 343–358.

Kolibri 360 GmbH
Ferdinand-Jühlke-Straße 7
99095 Erfurt, Deutschland
produktsicherheit@kolibri360.de